EARLY CHILDHOOD EDUCATION, POSTCOLONIAL THEORY, AND TEACHING PRACTICES IN INDIA

BALANCING VYGOTSKY AND THE VEDA

AMITA GUPTA

The extract on page 35 from Ignited Minds: *Unleashing the Power within India* by A.P.J. Abdul Kalam is reproduced courtesy of the publishers (Penguin Books India Pvt. Ltd.) and the author.

Figure 6.1 on page 119 has been reproduced with the permission of Oxford University Press India, New Delhi.

The extract on page 75 has been reproduced with the permission of Oxford University Press India, New Delhi.

The extract on page 23 has been reproduced with permission from Jeffery Paine, author of *Father India, Re-Enchantment, and Adventures with the Buddha.*

EARLY CHILDHOOD EDUCATION, POSTCOLONIAL THEORY, AND TEACHING PRACTICES IN INDIA
© Amita Gupta, 2006.
Softcover reprint of the hardcover 1st edition 2006 978-1-4039-7114-2

First published in 2006 by
PALGRAVE MACMILLAN™
175 Fifth Avenue, New York, N.Y. 10010 and
Houndmills, Basingstoke, Hampshire, England RG21 6XS
Companies and representatives throughout the world.

PALGRAVE MACMILLAN is the global academic imprint of the Palgrave Macmillan division of St. Martin's Press, LLC and of Palgrave Macmillan Ltd. Macmillan® is a registered trademark in the United States, United Kingdom and other countries. Palgrave is a registered trademark in the European Union and other countries.

ISBN 978-1-349-53282-7   ISBN 978-0-312-37634-5 (eBook)
DOI 10.1057/9780312376345

Library of Congress Cataloging-in-Publication Data

Gupta, Amita, 1959–
    Early childhood education, postcolonial theory, and teaching practices in India : balancing Vygotsky and the *Veda* / Amita Gupta.
        p. cm.
    Based on the author's doctoral research.
    Includes bibliographical references and index.
        1. Education, Primary—Social aspects—India. 2. Constructivism (Education) 3. Postcolonialism. I. Title.

LB1507.G85 2006
372.210954—dc22   2005053510

A catalogue record for this book is available from the British Library.

Design by Newgen Imaging Systems (P) Ltd., Chennai, India.

First edition: April 2006

10 9 8 7 6 5 4 3 2 1

Transferred to Digital Printing in 2008

Contents

Figures

Acknowledgments

Writing this book has been a tremendously exciting and rewarding journey for me and I want to acknowledge those who made this journey possible.

My mentor and advisor Leslie R. Williams at Teachers College, Columbia University, for guiding and facilitating my growth as a student and teacher. She, as a true *Guru*, encouraged me to formulate my own questions, map out my own quest, and seek the answers that would help me understand and contextualize my professional and personal beliefs, practices, and identity.

To the research participants in New Delhi who opened their hearts, homes, and classrooms to me, my sincere thanks; especially to those at "The School," for giving me the opportunity to learn about them as educators, and for letting me use their voices to share their thoughts and ideas with the world.

My professional colleagues and friends in New Delhi without whose constant encouragement, generosity of spirit, and ready willingness to help this project would not have been as smooth or as pleasant. In particular, I wish to thank Shayama Chona who took me under her wing when I needed help the most and opened the door to teaching for me; and Annapoorna Sehgal for her help and 40 years of friendship.

At Palgrave, Amanda Johnson for her encouragement and belief in my work; Elizabeth Sabo for seeing the book through production and for being patient with me while I tried to get the photograph for the cover just right; Emily Leithauser; and my copyeditor, Asha Boaz Ramesh for doing such a splendid job.

To many others from the various aspects of my life for their participation, support, and encouragement during the writing of this book: Afzal Friese, Amita Chauhan, Anita Malhotra, Asha Singh, Celia Genishi, Cheryl Ann Michael, Chitra and Raghav Sahgal, Deborah Flynn, Dev Pillay, Dolly Wadehra, Julia Sloan and Oleg Smirnoff, Nargis Panchapakesan, Renu and Nitin Joglekar, Sandeep and Padma Saxena, Shyamal Sanyal, Tulika Malik, Virinder Puri, Vijaya Hooja, and my colleagues at The City College of New York.

My grandparents and the invaluable lessons I learned from them—Aji who encouraged me to dream big and reach for the highest star; Baba who modeled for me an incredible love for poetry and a wondrous thirst for philosophy; Bauaji who taught me to be gentle and caring; and Babuji, who, at 98, continues to amaze me with his intellectual energy and passion for knowing.

My parents, Usha and Harimohan, for showing me how to combine career with family, for always believing that I could do it, and for being there through the highest crests and the deepest troughs in my life.

And most of all I want to thank my sons Nihaar and Naman for inspiring me and sharing in my excitement. Their unconditional love, positive energy, and valuable tech support enabled me to meet all the deadlines for this book. Nihaar with his generous, insightful, calm, and philosophical outlook, and Naman with his witty, organized, and practical approach have each brought only pride, joy, and sunshine into my life always.

This book is dedicated to my family who walks beside me, and to all the Saxenas and Rishis who have come before me and from whose collective wisdom I continue to learn.

Foreword

In picking up this book, you are about to begin a journey of discovery. If you are a teacher, you will find here clues to your own past, why you are the teacher that you are. And if you are a teacher educator, you have even more cause for celebration! Through her careful exploration, Amita Gupta will take you to the heart of one of the great conundrums of your work—lack of lasting effect of a program of teacher preparation on many of the teachers who have undertaken it. In spite of your best efforts, you may have been surprised, even disappointed to see your former students teaching in ways that are far from the ideals that you have cherished and were sure that you had promulgated.

In early childhood education, as in the field of education as a whole, one of the most persistent patterns in the work of teachers has been a separation between theory and practice, between what teachers were taught in their programs of preparation and what they actually do in their classrooms. While this has been a fact long observed, it has only recently been studied.[1] Teacher educators have been learning that there are many reasons for the lack of connection between practice and theory, foremost among them the deep-seated values and beliefs about learning and teaching that teachers bring with them into their classrooms from their own earliest recollections.[2] Although theories of child development and learning have been built on observations of children's growth and development within particular societies, and from examination of interactions between children and their teachers in those same contexts, little overt recognition has been given to the ways those descriptions have been influenced by the larger cultures in which such observations and interactions occurred. Much was assumed about the universality of descriptions of child development, and about the appropriateness of the recommended teaching strategies.[3]

Over the past two decades, awareness of the relevance of culture in general and of values and beliefs in particular for how teachers teach has grown among teacher educators.[4] In the United States, as in other societies greatly influenced by Western traditions of psychology and education, teacher education programs are being redesigned in ways intended to create a constant dialog between practice and theory, with particular attention to the power of the biographies of novice teachers[5] for the outcomes of teacher identity and formation. Now we are beginning to see clearly that who we are culturally is a major influence on who we become as teachers.

Whereas the issue of dissonance between who we are as people within a culture and who we are expected to be as teachers in our classrooms has been felt by many teachers and teacher educators in Western societies, perhaps nowhere has it been felt more strongly than by educators in postcolonial societies around the world. In countries such as India with more than 5,000 years of rich tradition in the human arts and sciences, the experience of colonization by other countries whose traditions were at extreme variance with their own was onerous across all the dimensions of human existence. It was in the area of education, however, that the efforts of colonizing powers were concentrated, in order to secure the acculturation of future generations and, thus, to continue domination by the colonizing state.

In India as one example, the education systems that had previously existed were replaced by systems that reproduced those of the colonial powers, and both the content and the processes of education became increasingly detached from the everyday realities of Indian life. Indian teachers likewise were prepared in manners similar to that of teachers in the colonizers' home countries; and as knowledge in the Western social and educational sciences grew, paradigms from fields such as child development and learning theory were introduced into the preparation of Indian teachers as well. With each addition has come a further distancing of Indian teachers from the formal systems of thought that had been part of their own cultures.[6]

Gupta points out that as this process was going on, children in Indian society were being raised in families who were enacting a way of life drawn from ancient sources—a way of life that had not and could not be erased by the experience of colonization. The tacit allegiance of the Indian peoples to the values and beliefs that had sustained them for generations did not change in its most fundamental manifestations. In her work, Gupta reveals the sources of these values and beliefs and illustrates the many ways they have continued to be woven through the rearing of children to the present day. These were the values and beliefs brought by novice teachers to their programs of preparation, where they encountered theories of a different sort, embodying values and beliefs very different from those that had shaped their own consciousness. These Indian teachers-to-be learned their lessons well. They were able to pass their teaching exams beautifully. But once in their own classrooms, they again (like teachers in other parts of the world who have experienced such dissonance)[7] reverted to teaching as an expression of their deepest leadings.

The wonder of Gupta's account of this process is that it can be read on several different levels at once. Certainly teachers in India today will resonate to her descriptions, recognizing their experience in her own. I have found as a Western teacher educator, however, that I also recognized her description as something that happens with the students whom I am preparing to be teachers, regardless of their particular cultural background. I work in the United States, but still I see a great separation between the values and beliefs of many of my students and those that are embedded in the theories that are expected to undergird our teaching practices. Reading Gupta's description of the process in a cultural context different from my own has allowed me to see more clearly the dynamics of my own situation—something that familiarity had obscured. Thus, I have learned from reading this book, and I have been inspired to renew my efforts to bring into play students' values and beliefs as they expand their repertoires and learn other ways to see the children they teach. I have been strengthened in my understanding that an effective future builds on a recognized past, and that the present moment is the time to work toward a wholeness of experience in both teaching and learning. I wish you the same pleasure in Gupta's work.

Leslie R. Williams, Ed.D.
Professor of Education
Teachers College, Columbia University

Notes

1. L.R. Williams (1996), Does practice lead theory? Teachers' constructs about teaching: Bottom-up perspectives, in S. Reifel & J.A. Chafel (eds.), *Advances in early education and day care: Theory and practice in early childhood teaching*, Vol. 8 (Greenwich, CT: Jai Press).
2. D.F. Brown & T.D. Rose (1995), Self-reported classroom impact of teachers' theories about learning and obstacles to implementation. *Action in Teacher Education*, 27(1), 20–29; Spodek, B. (1988). Implicit theories of early childhood teachers: Foundations for professional behavior, in B. Spodek, O.N. Saracho, & D.L. Peters (eds.), *Professionalism and the early childhood practitioner* (New York, NY: Teachers College Press), pp. 161–172.
3. S. Bredekamp & C. Copple (eds.) (1997), *Developmentally appropriate practice in early childhood programs*, rev. ed. (Washington, DC: National Association for the Education of Young Children); P.G. Ramsey & L.R. Williams (2003), *Multicultural education: A source book* (New York, NY: Routledge).
4. S. Grieshaber & G.S. Cannella (2001), *Embracing identities in early childhood education: Diversity and possibilities* (New York, NY: Teacher College Press).

5. B.T. Bowman (1989), Self-reflection as an element of professionalism, in F. O'C. Rust & L.R Williams (eds.), *The care and education of young children: Expanding contexts, sharpening focus* (New York, NY: Teachers College Press), pp. 108–115.

6. I recognize that India, in its enormous complexity, has always had multiple cultures, and, thus multiple cultural influences upon its evolution. One cannot characterize India as a cultural monolith any more than one can see Western societies in that way. For that reason, I speak of cultures and colonizers in the plural, understanding that the interactions to which I allude are equally complex.

7. In my own work with teachers in several different countries, I have found a similar separation to be present between their practices and the theories they had studied during their preparation as teachers.

Preface

> *. . . we must abandon the naive idea that facts exist by themselves, universally recognizable by anyone who cares to look. The fact is that no fact is a fact unless the structure and context which gave it birth are discovered simultaneously with the fact. Facts are facts because they come in a web of conceptual structures.*
>
> —de Nicolas, 1976, p. 186

The seeds for this book were generated within the sociocultural context of my life. As a young girl in school I was an avid reader of fiction, fascinated by books on international espionage, and I had always fantasized becoming a submarine captain when I grew up (yes, I had read one too many Alistair MacLean thrillers under my mother's influence). However, belonging to a family in which every second adult was a physician, there was a general expectation that I would also go to medical school and become a doctor. In college, I was intensely focused on the study of my undergraduate science major coursework and passed my annual essay-based examinations with honors, ranking second in the Delhi University. But when the time came to take the examination for being admitted into medical school, I had not practiced enough sample tests and had failed to develop the skills to ace the standardized, multiple-choice format of this exam. The result was that I did not get into medical school.

I then let my family know that I was agreeable to the idea of getting married and happily left the marriage-arranging decisions to them. And so it was arranged that I would marry into a family who were friends with friends of my family. A long engagement followed and to bide my time I applied to study for a B.Ed. degree, which would take nine months to complete, the exact duration of time as my engagement. Immediately after graduation, I was married and began a life of a full-time wife and stay-at-home mom. Due to major issues of incompatibility my marriage ended in a divorce eight years later, and I found myself as a single mom with the sole custody of two wonderful sons, on a desperate job search for the first time in my life. I needed an income to care for my sons and the first job I was offered was

as a nursery school teacher because I had happened to have a degree in education. I remain eternally grateful to the principal of that school who gave
me my first break. Thus began my journey into the world of early childhood education just at the time when my own sons were three and six years
old. I taught nursery school for four years, feeling a growing fascination for
the subject and an increasing desire to deepen my theoretical understanding
of child development and early education. My family, as always, encouraged
me to follow my desire and due to the unavailability of a higher degree program in the field of early education in India I applied to graduate schools in
the United States. The next several years found me in New York City at
Columbia University's Teachers College pursuing a Masters and then a
Doctoral degree in Early Childhood Education. Having to care for my sons,
I also found my first job as an assistant teacher in a progressive
Nursery/Kindergarten school on Manhattan's Upper West Side.

This interfacing of theory and practice in my work and study was the
perfect medium within which my own curiosity was nurtured and my
bicultural teaching experience began to confront questions in "developmentally appropriate" issues to which I had no answers at that time: Why
do young children in schools in India have to sit at desks? Why are there so
many children in each classroom? Why do teachers teach children academic
skills at such a young age? How come there is no social-emotional curriculum in the early childhood classrooms? Why do teachers 'tell' children what
to do thus impeding the latter's creativity? Why do teachers expect children
to use pencils at the age of three years? All these questions seemed indirect
criticisms of the educational approach in Indian schools, and none of my
colleagues ever asked me then in 1993 why Indian students in general performed well in their academic work. I would get the impression that strong
academic performance and the preparation of skills leading up to that were
not considered to be important in the early years of schooling. (And of
course nobody could have predicted that by 2005 India would emerge as a
global technological superpower because of the value given to academic
excellence especially in math and science.)

After having been introduced to the theoretical ideas of Piaget, Erikson,
Skinner, Dewey, and the philosophy of progressive education in the
stimulating classrooms of Teachers College, I too, at first, questioned the
approach to early education in India and found many of the elements
inappropriate. I had no answers at that time to the above questions. But
somewhere in the churning of those questions other musings began to crys-
tallize. The fact that startled me most was if early education in India was so
focused on academic teaching that there was little scope to address the
social-emotional development of young children, how was it that children
in India did possess reasonably competent social and emotional skills? In

fact, in my own teaching experience over the years, I had found most students in my classrooms in India to be quite emotionally stable, sociable, caring, responsive to rules, resilient, and enthusiastic about their teachers and friends at school. And my own small-sized progressive classrooms in the United States, in which the curriculum was strongly focused on socio-emotional development, were not necessarily lacking in children who exhibited socio-emotional concerns.

That must have been when the topic of my inquiry became clear to me. I wanted to have the appropriate answers to all of those questions. I wanted to be able to articulate why and how early childhood educational approaches could differ so greatly and yet find degrees of success in individual societies. I wanted to know whether and how the social and emotional skills prioritized in the United States would be equally prioritized in Indian society. And I wanted to understand why the criteria that characterized good teaching were so vastly different in the two countries. These fundamental and nagging questions, along with an advisor who was incredibly encouraging and supportive of my inquiry, found me on the path of an amazing research process.

Chapter 1 locates the book within urban India, problematizes the tensions and the teaching and learning processes, and briefly discusses the theoretical frameworks for this book, namely sociocultural–historical constructivism and postcolonialism. This overview of the theoretical frameworks at the beginning of the book will allow the readers to situate the findings of the study and subsequent discussions within the sociocultural and postcolonial context of the book.

Chapter 2 provides an overview to address the nature of Indian philosophy, and describes briefly the basic tenets of this philosophy, as well as the ancient texts and scriptures that serve to define this philosophy such as the *Veda*, *Upanishad*, and *Bhagvad Gita*. A short section on the multiple dimensions of diversity found in India is included to remind readers about the levels of complexities that have to be considered in order to navigate Indian society.

Chapter 3 provides: a brief history of education in India, a historical overview of the influences that seeped into the educational system in India, such as Buddhist ideas, Islamic ideas, and European initiatives; the structure of the current system of schooling; and an outline of some of the major early childhood programs currently available in India.

Chapter 4 presents some of the key findings from my research with regard to the aims of education for young children as considered to be the most important by educators in India. Classroom observations I conducted revealed that these aims were not only merely professed by teachers but also actively practiced by them in their classrooms. A detailed discussion of the sociocultural context within which these educational goals are embedded is presented. Clear connections are made between the concepts of *dharma* and *karma* as practiced in daily life and in an educational context.

In chapter 5, I use the sociocultural context of India to describe how the image of the teacher is culture bound. The discussion addresses how these images play a key role in defining appropriate adult–child relationships in any society. I also draw upon research data to make some comparisons between the role and responsibilities of the teacher in India and the United States, and discuss pedagogical freedom and choice in the classroom as a matter of perception and relativity.

The extremely important but often overlooked fact of why and how educational goals for young children can be different in different societies is discussed in chapter 6. I emphasize the varying significance attached to developmental milestones in different cultural societies, such as India and the United States, and present an in-depth description of how this significance determines educational goals for young children in each of the two school systems. Clear examples of appropriate developmental and social norms as well as educational aims are provided.

In chapter 7, I describe what, in the teachers' opinions, were the strongest influences on their learning to become teachers by including powerful and insightful narratives and voices of early childhood teachers. I have connected these particular findings to a process of sociocultural constructivism in the teachers' own learning. I also trace the intersections of what teachers learn from their sociocultural experiences and what they learn from their formal teacher education programs to the overarching educational aims and their own classroom practice.

Chapter 8 focuses on the issue of class or group size, particularly in the sociocultural context of schooling in India. I discuss data from my research that shed light upon the teachers' perspectives on working with a large number of students in one class. I describe how teachers with an average class size of 40 students implemented their curriculum and classroom activities. I also list specific indicators of successful teaching that were discerned from the research data, and contrast the different learning styles and modes of engagement that children demonstrate in an Indian classroom versus a progressive classroom in the United States.

Chapter 9 includes a description of the early childhood curriculum that I found being implemented in the classrooms of three teachers, and I discuss the sociocultural nature of this curriculum paying special attention to how three culturally different discourses interface within a postcolonial context.

Chapter 10 includes a short discussion on some of the major teacher education approaches that have been adopted by Euro-American colleges of education, and a short description of the teacher education environment and curriculum in New Delhi. I then present an argument for an appropriate teacher preparation experience for teachers in India whose lifestyles and teaching practices already reflect a postcolonial hybridization of both traditional Indian and Euro-American ideas. I discuss also whether this attempt would empower Indian teachers to participate in an international early education discourse, or it would marginalize them even further in the arena of that discourse.

Chapter 11 is a reflective journal in which I share the experiences, tensions, and biases that I felt while I traversed the path of academic research in a postcolonial context.

I hope that this book provides an in-depth understanding of how school success is defined in Indian society, and what sociocultural factors make the early childhood approach in India so different from the standard dominant discourse of early childhood education in the United States. I want to be particularly careful about not generalizing and giving the reader a simplistic view of the educational system in India. India with all her religious, ethnic, socioeconomic, linguistic diversity reflects the same degree of diversity in her schooling systems. This book is an account of a narrow section of education—the urban, private school, early childhood practice—and certainly does not represent all early childhood classrooms in the country.

Chapter 1

Conceptualizing and Setting the Stage

Although extensive research has been conducted on early childhood education in India, most of it has been focused on initiatives and projects sponsored by the Indian government centered on rural areas or urban slums for children from low socioeconomic and rural communities. Or else it has been sponsored by nongovernmental organizations (NGOs) and international organizations such as the UNICEF and UNESCO as well as the World Bank, but which have also focused mostly on educational development at the grass roots level. The main thrust of these programs has been on researching and developing projects to ensure that children living in rural and economically disadvantaged communities get basic safety, hygiene, and nutrition. Consequently, there has been almost no in-depth formal research conducted on early childhood education in the private urban schools with which the middle-class Indian family is associated. The rapidly growing Indian middle-class population of almost 350 million (more than the total U.S. population) includes lower-middle-class and upper-middle-class families belonging to social and occupational groups such as those employed in the government service; doctors, engineers, lawyers and other such professionals; well-to-do merchant families in the business and trading occupations; school teachers in large urban centers and in institutes of higher education; those in journalism and media; those who are partially or fully educated among the middle-level peasantry; the white-collar workers in the private sectors; policy makers and legislators; and a considerable population of university students. It is estimated that by the year 2025 this middle-class population will total 550 million people (Varma, 1999). My objective is to bring the Indian cultural and postcolonial perspective of early childhood teacher preparation and practice to the discourse of global early childhood education. This perspective may help shed light on aspects of urban schooling in India, such as the nature of a socially constructed early childhood curriculum; early childhood pedagogy, teachers' intrinsic beliefs, and the tensions between teacher preparation and practice; the challenge of working with large classes; the prevailing image of the teacher and child in Indian society; understanding the measures of school success for young children; and so forth. I would like to emphasize that the ensuing discussions are not, by any chance, reflective of all of the schools that exist in India—a hugely diverse country in terms of religion, ethnicity, language, socioeconomic classes, and the stratified caste system. This book is based on the study of a small slice of the much larger system of education in India, but it is an important slice that has not been researched enough.

I began my study with certain assumptions, starting with the notion that early childhood education in India was informally influenced by local intrinsic beliefs, values, and expectations, and formally influenced by the Euro-American discourse of education that dominates the field in general. Starting early in life, and through informal and indirect channels, Indian children begin to learn the values inherent in India's cultural philosophy, which is based primarily on Hinduism. My second assumption was that teachers in urban India faced a cultural discrepancy between the educational philosophy upon which their professional training was based and the Indian philosophy upon which their actual classroom work was based. The values and beliefs that teachers learned through their social and cultural experiences influenced their practice to a large extent. My third assumption was that the lifestyles of people in urban India are based on their local and traditional belief systems, as well as on the more modern ways of life as seen in younger societies of the West. My fourth assumption was that my research would provide the space for the voice of the "other" to be heard, the "other" being the marginalized, non-Western early childhood teacher who strives to be the "right" teacher, who feels pressured and compelled to follow the standards of early childhood education that have been articulated within the early childhood discourse that is dominant in the "West"—the "West" meaning English-speaking, predominantly "white" nations such as the United States, the United Kingdom, Canada, and Australia.

Urban India in the context of this book refers not only to the large metropolitan cities in India but also to smaller cities. Let me attempt to provide a snapshot image of the city of New Delhi, where I conducted my study. New Delhi is the capital of India and is the seat of the central government. Located in northern India, it is one of the four largest metropolises in the country with a population of about 13 million people spread over an area of approximately 1,500 square kilometers. The city is a busy, congested mix of residential areas reflecting both palatial mansions as well as poor hutments; commercial districts that combine large department stores, private offices in high rise buildings, and also small roadside stalls; government offices situated in colonial as well as modern buildings; broad, sweeping flyovers and freeways but also dirty, narrow alleyways; ensnaring billboards depicting clever advertisements, colorful products, and places, as well as garishly painted famous Bollywood heroes and heroines; countless sprawling bazaars and marketplaces where one can buy anything from small-scale handicrafts indigenous of any state in India to Louis Vuitton luggage; countless educational institutions from nursery schools to colleges and universities; places of worship on almost every street, which include Hindu temples, Muslim mosques, Christian or Catholic churches, Sikh gurudwaras, and so forth; countless food shops and restaurants reflecting cuisines from all over India and the world; millions of people speaking in hundreds of languages; vehicles that outnumber the combined total of vehicles in the three other metropolitan cities of Mumbai, Chennai, and Kolkata; hundreds of thousands of phone booths, Internet cafes, and call centers for multinational corporations; and always the incessant noise of traffic, people, and music. Although I have not made even a dent in describing the complexity of the city of Delhi, I hope that these few sentences provide the reader with some sense of the environment within which this study and its participants were situated.

Problematizing the Status Quo in Teaching and Learning

In light of the diverse kinds of early childhood schools found in India, which have been described in more detail in chapter 3, and in order to keep the research manageable, my study included a very specific slice of Indian education—the urban, secular, private, early childhood/elementary dimension. The reason was because I could better compare and contrast issues of early childhood educational practice between such schools in India and schools in the United States; these two school systems were more comparable in terms of the resources available, teachers' educational levels and qualifications, and resources accessible to students in their home environments.

A review of randomly selected mission statements and brochures of private schools in New Delhi revealed that the articulation of the institutions' educational goals and school philosophies was couched in the language of the philosophical texts indigenous to India as well as in the language of

Western progressive early childhood education. One school described its progressive approach and commitment to excellence, and a curriculum that incorporates the principles of *Jeevan Vigyan* (the science of living) such as meditation, relaxation, time management, and yoga. Another school was described as a coeducational, English-medium school with an emphasis on strong Indian cultural values, national integration, and brotherhood, as well as on holistic education to enhance children's cognitive and noncognitive skills. Yet another school claimed to be progressive, child-centered, and coeducational, fostering creativity, independent thinking, and exploration, as well as sound ethical values such as kindness, respect, and integrity. A fourth school offered an educational system based on universal values, excellence in all things academic as well as nonacademic, global understanding, valuing diversity, and service to humanity. Many of the schools also have names and school mottos that reflect Sanskrit words and phrases taken from Hindu scriptures and philosophy.

At the same time, an analysis of some teacher educational programs in New Delhi (University of Delhi 1999–2002 scheme of examinations and course readings for degree study in elementary education and home science; the State Council of Educational Research and Training (SCERT) curriculum for elementary teacher education for 1999) revealed a curriculum that was largely devoted to topics in Western educational and developmental psychology; pedagogical methods and materials; and the teaching of content areas. Child development appeared as a core course, but the course description drew mainly on Euro-American theories of child psychology. Current and updated early childhood theory and research as well as cross-cultural perspectives on child development seemed to be missing. Also missing was a comprehensive study of the history of Indian education, and the recognition of the existing connections between Indian philosophy and the educational curriculum in Indian schools. The history of educational reforms and policies was addressed, but was limited to periods in India's history occurring slightly before and immediately after her independence in 1947. With regard to the sociocultural context, very little was included in the overall teacher education curriculum about the philosophy that forms the basis of the Indian worldview and that constitutes the Indian-ness that is characteristic of the diverse people of India.

Within my own bicultural experience of teaching and supervising early childhood educational centers, I had learned how to be a "good" teacher in a classroom in New Delhi as well as in a progressive classroom in New York, and I was very aware that my teaching objectives, dispositions, demeanor, strategies, and approaches were vastly different in the two culturally different situations. Good teaching is not thought of in the same way across different cultural communities (Delpit, 1995), and this was certainly true in how teaching and learning were viewed in the early childhood progressive classrooms in New York and in the early childhood private classrooms in New Delhi. Such observations recorded by earlier researchers on the pedagogical conflicts emerging due to cultural differences in classrooms with multicultural personal and pedagogical belief systems led to several important studies concluding that developmentally appropriate practices in early childhood education, based largely on Western child development theories, were inappropriate and inapplicable across several non-Western cultures and communities (Bernhard, Gonzalez-Mena, Chang, O'Loughlin, Eggers-Pierola, et al., 1998; Jipson, 1991; Katz, 1996). Child development is a very Euro-Western concept and emphasizes the individualistic nature of optimal development (New & Mallory, 1994). Although all children have biological propensities for growth and development, the significance of their developmental milestones and behaviors is determined only by the values and expectations of their cultures (Bowman & Stott, 1994). Early childhood discourse has been widely defined by child development theory, so much so that it has prevented educators viewing it from any alternative perspective considering those perspectives as unscientific and inaccurate (Bloch, 1992). The appropriateness of teachers' classroom practices has thus been based upon a set of developmental skills that were deemed appropriate and acceptable primarily in the white middle-class communities of Europe and the United States. The initial manual of *Developmentally Appropriate Practices* by Bredekamp (1987), endorsed by the National Association for the Education of Young Children (NAEYC), was subsequently revised in 1997 to reflect teaching guidelines that would be more culturally appropriate (Bredekamp & Copple, 1997). Many critics feel that even the revised edition has not addressed the full extent of this issue (Grieshaber & Canella, 2001), and much of early childhood practice in the United States still follows the momentum set in place by the original version after it was endorsed by a renowned national organization such as the NAEYC. Having been trained and having taught in early childhood classrooms both in India and the United States, I have myself come face to face with some of the tensions between Western child development theory and the Indian value system, which differ in that different developmental milestones and skills in children are given more significance and importance. I briefly introduce some of these contradictions here but have provided a more detailed discussion on the appropriateness of social norms and developmental skills for young children in the context of Indian society in chapter 6.

One of the basic and most prominent cultural differences between the Indian and American worldviews is the concept of the "self." In India, the development of the self is important, but the self is typically viewed in a social context (Dave, 1991; Kakar, 1981), and more often the "non-self" or the "other" is given more importance than the "self." In an Indian early childhood classroom, a child playing with a toy would more likely be expected to share the toy when approached by another child rather than keeping it to himself/herself. Acting on the right of the individual not to share would definitely be viewed quite negatively as being selfish and unsocial in this instance.

Healthy social development within the tradition of progressive education in the United States has been defined within the concepts of leadership skills and public speaking abilities. Verbal participation, oral presentations, and articulation of ideas have been a primary objective, and the more articulately an individual can "sell" his/her skills and accomplishments the more assured he/she is of social success. The interview is of critical importance in the process of applying for college admissions, jobs, and internships. Children in schools are thus encouraged to develop these verbal abilities at a young age, and the acquisition of these skills becomes an important criterion in assessing children's social development. The Indian worldview recognizes the complexities of emotions and the inadequacy of language to convey these complexities. College admissions, job offers, and promotions in India are more dependent on the applicant's actual academic performance and accomplishments. When viewed through the lens of Indian culture, a lens that gives more importance to the "other" rather than the "self," self-praise is considered to be arrogant and it is up to others, more specifically supervisors, to value and assess an individual's work. In India, healthy social development has been viewed typically more in terms of group dynamics and developing a sense of duty and responsibility toward the group. It reaches beyond verbalizing and emphasizes the actual completion of a task to measure an individual's competence.

Another pedagogical conflict is in reference to the concept of focus in young children. The individual's ability to focus is valued in both systems, but in India, the early development of this skill is facilitated when young children engage in activities that demand a longer attention span, attention to small detail, the use of fine motor skills, and waiting quietly and patiently when it is required of them. This is in contradiction to the recommended developmentally appropriate practices based on Euro-American child development theories, where young children are encouraged to physically move not sit, and use larger tools and manipulatives rather than finer ones. Even the nature of physical and emotional development is recognized variously in different cultural contexts. There is no hurry to help the three- or four-year-old child to become as independent as possible as early as possible. The more valued developmental goal in India is interdependence rather than independence.

By and large, the social and emotional development of children in India occurs in an on going manner outside of school. Thus traditionally, the role of the teachers and the purpose of the school's formal curriculum have been to focus more on the teaching of academics. Academic teaching was not substituting for, but was adding to, the social learning that was already happening in family and community contexts. In the Indian understanding of the teacher–student relationship, the teacher is considered to be the wiser and more experienced adult who is expected to guide the student in the right direction rather than adopt a stance of noninterference lest it impede the child's curiosity or initiative as was recommended by some learning and developmental theories in the West.

Questioning the existence of universal patterns of child development, and arguing that development is a function of culture, it has been strongly recommended that each society identify appropriate and inappropriate practices based on values that are prioritized by different cultural groups (Bernhard et al., 1998). Socio-emotional, cognitive, and motor skills that are prioritized by members of a given cultural community or society get more supported and nurtured in classrooms as well as in homes and thus become more prominently developed within the given community. In chapter 6, I provide further details on this recommendation in an attempt to identify those skills that are regarded as being more important for children growing up and succeeding in Indian society.

Learning as a Sociocultural–Historical Constructivist Process

Traditionally, dominant child developmental theories in the West have viewed individuals as physically constructing cognitive models independently, without recognizing the importance of social and cultural processes at work. Thus rational constructivists, led by Piaget, consider the child as an egocentric self and the construction of knowledge as an individual and cognitive process. The focus of Piaget's research has been on genetic epistemology and its scientific discourse has described the human being in such clinical terms as individual, organism, and biological (Soto, 1999). Piaget's theory explains cognitive development and learning in the human being in terms of the physical interactions between a biological organism and a physical environment through a process of assimilation and accommodation. This precludes the influence that social, cultural, and historical factors—by way of interactions with people, language, traditions, and rituals—have on children's learning and cognitive development. Piaget's emphasis on logicomathematical reasoning denies the multiple realities of children's worlds and the powerful role that fantasy plays in these realities. It is interesting to note that an emphasis on logico-mathematical reasoning is also reflected in the value placed on democracy, capitalism and free choice in American society, and the hope that people would use the power of rationality to make the right choices, a prerequisite being that many choices must first be made available. Thus reasoning and thinking in a logical and independent manner become the desired and necessary skills that would allow an individual to succeed in American society. This subsequently identifies the behaviors and responses that would be considered as socially acceptable in this society, such as independent decision making, risk-taking, and so forth. It is in the prioritizing and selecting of acceptable behaviors in the course of children's growth and development that the difference between cultures and communities becomes clearly evident. Historically, developmental theories such as that of Piaget have been given a prominent place in early childhood teacher education programs. According to the initial version of the NAEYC guidelines set forth in the United States, Piaget's theory lays down one of the most important foundations for *Developmentally Appropriate Practices*, which in turn define and drive much of early childhood education (Bredekamp, 1987). This body of child development theory that relies heavily on Piagetian ideas is seen in the teacher education curricula in India also.

However, the worldview of the Indian people is highly social in nature, and childrearing in India is very much a social activity. Knowledge construction is viewed more as a social rather than physical process that draws on cumulative historical and cultural experiences in the individual's past and present. Human thought has been described to be essentially social in its origins, functions, forms, and applications, and thinking as largely a public activity, occurring spontaneously and naturally in the yard, marketplace, or the town square (Geertz, 1973). This description certainly characterizes the social nature of the Indian way of life and thought. As a socio-constructivist, Vygotsky too viewed the construction of knowledge to be not only cognitive but also social in nature with the belief that higher psychological development happens first at the social level and then is internalized at the individual cognitive level (Vygotsky, 1962, 1978). As individuals we attempt to reflect upon and thus generalize meaning across our experiences, and the ability to represent allows us to consider multiple perspectives simultaneously; further, meanings and worldviews, although unique to the cognizing individual, are the result of shared meanings by a community and are linked to culture right from the start (Fosnot, 1996). Meaning making, or ways of thinking and interpreting the world is thus closely related to social interactions and the language of these interactions.

The knowledge constructed in this manner is then transferred to children through dialogues with them and others. Thus language plays a critically important role in the construction of knowledge and meaning. It is only the recently revised 1997 version of *Developmentally Appropriate Practices* that recognizes somewhat the Vygotskian concept of the social nature of learning (Bredekamp & Copple, 1997).

The powerful Vygotskian notion of "the historical child" (Vygotsky, 1934/1987) supports the prime consideration of this book, which is to situate my inquiry within the particular cultural and historical conditions that provide the context for social views on how children are raised and educated in India. This notion also provided me with a context to explore how teachers may be further empowered to implement their practice so as to meet the educational objectives set out by schools in India. Thus sociocultural—historical constructivism as a framework for my study allowed me to shift the focus from individual students to the context in which the students were situated. On another level, societies are systems and to understand the actions of any part of this system (such as early childhood teacher preparation and practice) it became necessary to understand the unit as a whole, that is the pervasiveness of an ancient lived philosophy in Indian society. Since the research and discussion required me to move out from the local field of schools into a much larger world of context, it also became necessary for me to draw upon theories beyond those that I used to explain the local. Therefore, I have drawn upon postcolonial theory that allowed me to examine the interplay between the colonial and dominant discourse of education and the more local voices of education in India.

Locating the Study within a Postcolonial Discourse

Although the term "postcolonial" may literally refer to a historical period that marks the end of colonization and the beginning of political autonomy in a former colony such as India, it may be more appropriate to view it as the period of influence that began with the start of colonization. I view the term "postcolonialism" as a research paradigm that seeks to situate contemporary educational issues in the context of underlying colonial experiences. Postcolonial theory has highlighted the differences between the dominant and the marginalized, and has provided a platform for non-Western critics located in the West to present their cultural inheritance as knowledge. In particular I would like to recognize that what counts as marginal in the West is in fact dominant in the non-West (Gandhi, 1998), and that postcolonial study has been a critical examination of the past in an attempt to reveal "marginalized" experiences and an openness to multiple perspectives (Viruru, 2001). For the purpose of my discussions, I have engaged with the following definitions of the colonized condition: a powerful interdependence between the colonized and the colonizer (Gandhi, 1998); a continuing contest between the dominance of the colonizers and the consequent legacies that were created (de Alva, 1995); a transaction, a two-way dialogue between the philosophies of the colonized and the colonizer (Trivedi, 1993); a phenomenon of cultural hybridity (Bhabha, 1994); and the intercultural negotiation between the ideas of the colonizer and the colonized (Pratt, 1992). I have applied that premise to the field of education to examine the interactions between Indian educational philosophy and Euro/ American educational philosophies, the assumption being that the two are located in different worldviews: each perceiving the world differently.

The process of colonization in India began with physical conquest of territories followed by the conquest of minds, selves, and cultures. For Europe to be established as the site of civilization, the colonized world had to be emptied of its own knowledge and value. The West, thus, had to be established not only in structures but also in minds (Nandy, 1983). One of the tensions of the intercultural transaction was that although the colonized fervently disliked the colonizers, they also longed for the power and the knowledge that the colonizers possessed. The natives of India wanted the privileges and benefits of the English rule without the physical presence of the English people to rule them (Gandhi, 1938). To be in control, the colonizer had to essentially become the educator and proclaim that the only way colonized Indians could be properly educated was through the language and canon of the educator. In British India, this required the implementation of a system of education that would bureaucratically control the way natives would be educated. This assumption that proper education must be defined by the language and canon of the colonizer became the underlying basis for the famous Macaulay's Minute of 1835 in which the then viceroy of India, Lord Macaulay, proclaimed that the English language would become the medium of instruction in schools in India. The declaration of Macaulay's Minute was the beginning of a system of education in India that would tend to forsake its own history and philosophy and become Western. It influenced the passing of the English Education Act in 1835 requiring the natives of India to submit to its study (Viswanathan, 1995). In the previous section on sociocultural constructivism, I discussed briefly the prominent role language played in the construction of meaning. By making English the language of instruction and communication in India, meaning-making processes of the postcolonial Indian developed into a complex understanding of the world, which built on both Indian and Western ideas. This was to become a lasting legacy for the psyche of the modern urban Indian individual.

Another legacy that lingers from the British colonial era is the textbook-centered and examination-oriented pedagogy that began, and continues, to dominate education in India. This rote-learning system of education implemented by the British administrators in the early nineteenth century came to be governed by British bureaucracy that controlled all aspects of schooling. A third legacy of British colonialism was the creation of a class of native elites in its own image. The urgency of forming this elite Indian middle class was stressed by Lord Macaulay as he articulated that this group would serve as interpreters between the British rulers and the millions that they ruled, a class of persons who would be Indian in blood and color but English in taste and intellect (Varma, 1999). Thus these Indian Sahibs wore English clothes, spoke in English to each other, lived in bungalows furnished in the English style, observed English table manners, and basically tried to follow English etiquette (Husain, 1978). Often, however, they only succeeded in appearing uncomfortable and ridiculous. It was through such individual processes and lasting legacies that the intriguing transaction of the hybridization of colonizing and colonized ideas was established in the Indian mind.

At another level, the historical nature of the transactional relationship between Ancient Indian philosophy and Euro-American philosophers can be seen in the philosophical discourses of Heidegger, Nietzsche, Schopenhauer, Husserl, and others. These, and several other philosophers, were deeply influenced by encounters that grew out of the rise of Indology in the eighteenth century (Dhillon, 1997). The Western world's interest in the philosophy and scriptures of Ancient India began with the translation by Charles Wilkins in 1784 of Hinduism's most important text, the *Bhagvad Gita*. This directly influenced the works of not only the European philosophers mentioned above, but also that of writers such as Thoreau, Emerson, Whitman, E.M. Forster, and T.S. Eliot. It is thus evident that the transfer of ideas between Indian and Western philosophies began several hundred years ago.

The postcolonial discourse recognizes and addresses the exclusions in the Western canon and strives to make the field of knowledge more representational and inclusive. Based on her ethnographic study of an early childhood center in Hyderabad, Viruru (2001) puts forward the question of whether early childhood education can become something other than a field shaped by the dominant discourse. Much of this discourse is couched in the language of Western affluence and many elements of progressive early childhood education are remote from the worlds of children in the non-West. The earlier mentioned and greatly emphasized publication *Developmentally*

Appropriate Practices (Bredecamp & Copple, 1997) that defines American early childhood education highlights a set of skills and values for young children to learn that are appropriate for children growing up in middle-class America. The values, skills, and attitudes that are developmentally and socially appropriate for children growing up in India are quite different. A postcolonial framework enables the understanding of alternative perspectives by lending a ear to the intrinsic "other" voice of early childhood in a non-Western culture.

I have discussed the findings of my research and subsequent recommendations within the postcolonial framework of the complexities of cultural hybridization and a two-way dialogue between the philosophies of the colonized and the colonizer that bridges Euro-American and Indian thought. This concept of transaction and hybridization of ideas, language, ideologies, pedagogies, and curricular practices was clearly evident during the course of the research, specifically with regard to the nature of the school curriculum and the classroom practices of early childhood teachers who participated in the research. The manner in which three early childhood teacher participants in the study engaged in a negotiation between Ancient Indian philosophy, the educational philosophy of the British colonizers, and the more recent American progressive philosophy of education is described in more detail in chapter 9. However, although there was a hybridized quality to the psyche and classroom practice of the modern Indian urban private school teacher that was based on both Indian and Western ideas and life styles, the teacher education curriculum tended to focus more on the Euro-American psychologies and philosophies of education. It is one of the aims of my discussion to attempt to situate appropriately the educational philosophies of the colonizers and the colonized during the learning and practice of teachers in India, and subsequently encourage teacher education to draw on a matrix of histories and philosophies that represent traditional Indian ideas as well as Euro-American educational ideas. This is another aspect of "thinking our way through, and therefore out of, the historical imbalances and cultural inequalities produced by the colonial encounter" (Gandhi, 1998, p. 176).

Although postcolonial discourse runs both with and against academic discourse in the West, it still speaks in the voice of Western academia (Dhillon, 1997); if it hopes to bring about any change in the dominant discourse then the "other" discourse, in this case the local Indian philosophy, must be spoken of and its relationship to Western philosophy be established. Whereas a Western approach to philosophy largely emphasized science and logic, the senses, and materialism, the Eastern approach to philosophy, such as that of Indian, Chinese, Japanese, and Middle Eastern, emphasized the individual's inner world and intuition (Ozmon & Craver, 1995). Unlike

Greek philosophy, which separated philosophy and religion, in Indian school of thought philosophy and religion are closely and almost inseparably intertwined. Dhillon reminds us of two very significant moments in the relationship between Indian and European philosophy. The first is the separation of Indian thought into Philosophy and Religion, and the second is the process of erasure of Indian philosophy under the conditions of colonization and the hegemonies of Western society. The next chapter traces the nature of Indian philosophy, primarily Hindu philosophy, before this separation occurred.

I would like to emphasize that I am not valuing one educational approach over another in examining educational philosophies that originated in Europe and the United States and those that originated in Ancient India. Nor is it my intention, through this examination, to create additional binaries between "Indian" and "Western." I emphasize that there *is* a hybridization of colonizing and colonized educational ideas in the current Indian worldview, and my attempt is to contextualize these ideas so as to better understand the historical evolution of the worldview that forms the basis of the current system of education in India. In order to establish that that which is colonized and colonizing are not mutually exclusive in the mind of Indians, it is first necessary to begin to understand what is in fact Indian and what is Western. Further, I recognize the importance of the breakdown of the rigidity between binaries (Suleri, 1992). My aim is to describe the continuities and the constant flow of ideas and pedagogies that create a particular approach in early childhood education in many private urban Indian schools as became evident in this study. It was an approach that reflected values such as sense of duty, familial ties, and veneration of elders and ancestors, as well as ideas such as emphasis on material goods, social advancement, and changing standards, all coming together in a unique socioculturally constructed curriculum, which has been described in detail in chapter 9. In attempting to support the hybrid and transactional approach that seems to be in place in school curricula and teachers' own practices and beliefs at the early childhood levels, I am also urging the reexamination of the more rigid and colonial current teacher education curricula in India. In order for the latter to be more in consonance with the socio-constructivist hybrid nature of classroom practice, teacher educators need to explicitly acknowledge and explain to teachers how and why developmental and social norms for children growing up in India can be somewhat different from the developmentally appropriate ideals that drive early childhood education in American progressive classrooms.

Chapter 2

The Sociocultural Context of Education: Core Concepts of the Philosophy Underlying the Worldview of Indians

The discussions in the forthcoming chapters make frequent references to some of the core concepts of an ancient philosophy that forms the underlying basis of Indian society. I, therefore, present a short description of that philosophical and spiritual tradition. This chapter has been organized into sections and subsections that address relevant topics, concepts, and issues that characterize the Indian worldview and subsequent way of life. A brief overview of the scriptures and texts that define this philosophy is followed by a discussion on some of the underlying concepts such as the notions of self, knowledge and learning, moral development, and diversity as understood within the Indian worldview.

Indian Scriptures: Texts that Form the Basis of Philosophy

The philosophy of education in Ancient India was based primarily on the philosophical offerings of the scriptures namely the *Veda, Upanishad, Vedanta, Brahmana,* and the *Bhagvad Gita* of which the earliest are the *Veda*. In addition, much that is known of the beginnings of the 5,000 years of Indian civilization has been reconstructed from these written accounts. It is relevant to first become acquainted with some preliminary knowledge about these scriptures before attempting to understand Indian philosophy. Figure 2.1 graphically represents the chronological origin and evolution of Ancient Indian scriptures and major texts. Along with the name of each text is a brief description, with a more detailed summary of each scripture and text following the figure. It is not my intention in this book to critique these texts, or judge them to be right or wrong. My only aim is to familiarize the reader with the scope of some of the texts that continue to define the beliefs, values, and practices of a large percentage of the population in both urban and rural India.

Veda

The word *Veda* is derived from the root "*vid*," to know. It is said that in prehistoric times there were seven "*rishis*," or sages, teachers of humankind, who received the sacred knowledge that was revealed to them (Ellinger, 1995; Zaehner, 1966). This text was first handed down by word of mouth with a precision extending to emphasis, and this degree of preciseness in an oral tradition should not be underestimated. The first written accounts of the *Veda* can be traced back to 2000 B.C. Interest in the Sanskrit language began to increase in Europe after Charles Wilkins translated the *Bhagvad Gita* into English in 1784 (Basham, 1998). The German Sanskritist Friedrich Max Muller (1823–1900), who taught Comparative Theology at Oxford and provided one of the most splendid editions of the *Rig Veda*, the first book of the *Veda*, considered it to be the earliest historical account of world history (Chandra, 1980); the *Veda* are also considered to be among the oldest religious writings in the world (Keay, 2000). The language used for the earliest compositions is the archaic Vedic from which Sanskrit was later derived and is the root language of most European languages including Latin and Greek. The *Veda* consist of four books, and together the work is six times as long as the Bible. That the *Veda* were composed in northern India is supported by the fact that all the geographical limits mentioned in Vedic literature fall within the confines of India's original boundaries. The *Rig Veda* contains 1,017 hymns divided into 10 different books of which 7 form the original nucleus. Each of these 7 books is ascribed to a different seer and was passed down to his descendants. The books were thus family collections handed down from one generation to the next and were guarded jealously as a family inheritance (Keay, 1918/1980).

Vedic literature consists of a group of hymns, chants, and *mantras* to godly representations of three entities—earth, atmosphere, and heaven, with an attempt to bring about harmony between people's material needs

The major Ancient Indian texts and scriptures can be placed in two broad categories

SHRUTI	SMRITI
Writings that have a binding and canonical character, which are based on the sacred sounds heard by the ancient *rishis* or holy teachers. These texts include:	Scriptural writings emerging from what *rishis* experienced based on the *Shrutis*. Under this category are writings that include:

Veda: Consist of 4 books, which together are six times as long as the Bible. Said to have been written around 2000 B.C.
1. *Rig Veda*: Veda of the Verses
2. *Sama Veda*: Veda of the Songs
3. *Yajur Veda*: Veda of sacrificial sayings
4. *Atharva Veda*: Veda of mystical sayings

Samhita: These are a collection of songs and sacrificial sayings gathered from the *Veda*.

Brahmana: The introductions along with explanations from each of the *Veda*.

Upanishad: Philosophical and metaphysical insights collectively derived from all of the above writings and scriptures, written in about 800 B.C. They form the basis of the *Vedanta*.

Vedanta: Literally means the end of the *Veda*. These texts are composed of philosophical conclusions from the *Veda*.

Purana: Writings on creation, destruction, and new creation. Consist of 18 volumes containing a history of the universe, history of early Aryan people, life stories of Hindu deities. Written about the same time the *Veda* were written.

Tantra: Writings on the divine creative power and energy of the male principle *Shiva*, and the female principle *Shakti*. Although the notion of *tantra* in the West has been distorted and can have sexual connotations, it is the philosophy of union without which there can be no creation.

Laws of Manu: Written regulations and rules on social and political issues and questions, compiled in the early centuries A.D.

Darshana: The system of the six schools of Hindu philosophy: *Nyaya, Visheshika, Sankhya, Yoga, Purvamimamsa, Uttaramimamsa.*

Based on the philosophical conclusions of the above writings are the two great Indian epics that in themselves constitute two important sources of teaching and learning in all walks of life. They were initially written in about 1000 B.C. though other versions were also written later. These are the:

Ramayana: This is the older of the two epics, originally written in Sanskrit and consisting of 50,000 lines, and is the story of *King Rama* as one of the *avatars*, or reincarnated forms, of *Lord Vishnu*, his sacrifices in the name of duty and responsibility to his family, and the inevitable battle between good and evil with good triumphing in the end.

Mahabharata: Completed between 500 and 300 B.C., this epic conveys the story of conflict within a ruling family resulting in one of the mightiest battles ever fought, and includes the story of *Krishna*, the second *avatar* of *Lord Vishnu*. The *Mahabharata* includes the *Bhagvad Gita*, which is a philosophical song, dwelling on matters of duty, morality, and responsibility, and is considered as one of the most important guides to living a good life.

Figure 2.1   Origin and evolution of the major Ancient Indian texts and scriptures

and spiritual lives (Ozmon & Craver, 1995). Human beings are viewed as spirit, part of an ultimate source or reality, and religion consists of meditating on this spirit and leading a life of virtue and righteousness. The core ideas of the *Veda* permeate the other Hindu texts that followed and are integrally embedded in the philosophy, literature, poetry, the epic mythology, art, music, and dance of India.

Brahmana

These are works in prose and were composed between 800 B.C. and 500 B.C. They are connected with all the different Vedic schools. Besides instructions and explanations about sacrificial rituals, they also contain mythology, speculation, argument, and the beginnings of grammar, astronomy, etymology, philosophy, and law. By this time in history there had emerged a stratification of society into social and occupational groups. These groups or classes were recognized as *brahmans* (priests), *kshatriyas* (noblemen and warriors), *vaishyas* (traders and agriculturalists), and *sudras* (sanitation workers and artisans).

Upanishad

These are treatises dedicated to philosophical speculation and represent the last stage of the *Brahmana* literature. They are considered to be the teachings of a *guru*, or teacher, to the students. These were composed sometime between 800 B.C. and 500 B.C. The higher philosophical knowledge they offered came out of a sense of dissatisfaction from the ritualistic forms of worship that appeared in the *Veda*. Pupils having studied all the *Veda* and sacrificial rituals turned to the study of the *Upanishad* for knowledge of deeper philosophical speculations and answers. The *Upanishad* view life as marked by pain and suffering, which is the result of unfulfilled needs and desires, through hunger, thirst, sickness, frailty, deception, guilt, violence, and so forth. However, the *Upanishad* also recommend that the choice of being able to deliver oneself from this suffering is through one's own actions or *karma*. Deliverance can be achieved only by a true knowledge of the absolute or ultimate reality obtained through purity of life and meditation (Ozmon & Craver, 1995). The emphasis in the *Upanishad* is on the search for truth, and on knowledge of the individual self and the absolute self. The philosopher Schopenhauer considered the *Upanishadic* texts as being permeated by an earnest spirit, with original and sublime thoughts arising from every sentence; a study of these texts made for one of the most elevating experiences.

Vedanta

This term literally means the end, *"ant,"* the completion and crown of the *Veda*. The work is regarded as the concluding reflections on the *Veda*, emerging from the *Upanishad*. These writings essentially deal with the metaphysical relationship between the individual souls and the absolute soul, and the nonduality of the universe in terms of the absolute soul. It thus questions the nature of the material world and expresses it as being *maya*, an illusion.

Laws of Manu

It is said that Manu's Laws, books of rules and regulations, appeared in the first few centuries of the Christian era. These texts held great significance in the lives of Hindus and laid down codes that still influence social life and customs in India today. Most significantly, Manu was instrumental in defining the four castes that solidified into the controversial Indian caste system. He also specified the stages of life that a man and a woman should live by: youth, middle age, and old age.

Bhagvad Gita

Although the *Bhagvad Gita* is the account of just one episode in the great Indian epic *Mahabharata*, it stands complete in itself as 700 verses, and is described by William von Humboldt as the most beautiful, philosophical song existing in any known tongue (as cited in Nehru, 1946/1991). It describes a great war and the central theme dwells on the anguish that Arjuna, one of the main figures, feels when he realizes he has to fight his own beloved extended family to defend the larger society from moral collapse. Nehru describes it as a poem dealing not only with a political and social crisis, but also with a crisis in the spirit of man (Nehru, 1946/1991). The *Gita* deals with the spiritual background of human existence, and is a call to meet the obligations and duties of life while at the same time keeping in view this spiritual background and the larger purpose of the universe.

A modern renaissance of Hinduism, in the nineteenth and twentieth centuries, with a reiteration of its underlying concepts has been led by men such as Rabindra Nath Tagore, Sri Aurobindo, Dr. S. Radhakrishnan, Vivekananda, and Mahatma Gandhi, among many others.

Concepts Underlying Indian Philosophy

Indian philosophy is primarily Hindu philosophy and draws upon the Hindu texts described above. There are six recognized systems or schools in Hindu philosophy and they are *Nyaya, Vaisheshika, Sankhya, Yoga, Purvamimamsa,* and *Uttaramimamsa.* The base for all the schools is the same and together they provide an interpretation of what is the ultimate reality. Each school is based on the same metaphysical doctrine, and discusses one particular aspect of the whole (Bernard, 1995).

The first is a system of logic concerned with the means of acquiring right knowledge. The purpose is to enable us to attain salvation by classifying the different ways in which knowledge is acquired. This system or school is called *Nyaya.*

The second school or system classifies all knowledge of the objective world under nine realities: earth, water, fire, air, ether, time, space, soul, and mind. By having knowledge of Reality, objects that we perceive will no longer awaken within us any feeling of attraction or aversion, which is the source of all pain. This system is called *Vaisheshika.*

The third school of Hindu philosophy demonstrates that all derived experiences in this world are produced from two realities—Spirit and Matter. It views the evolution of matter from its cosmic cause as a projection of potentialities into realities according to fixed laws that can be understood and controlled by man. The fundamental tenet of this school is that creation is impossible, for something cannot come out of nothing. Whatever is, always is; and whatever is not, never is. This third system or school of philosophy is called *Sankhya.*

The fourth system is concerned with the ways and means by which the individual can know reality through direct experience. It is a system for perfecting human efficiency. The purpose is to free the individual from the three kinds of pain—those arising from one's own infirmities and wrong conduct; those arising from one's relationships with other living things; and those arising from one's relationship with natural and supernatural elements. This school of Hindu philosophy is known as *Yoga.*

The fifth system is concerned with the correct interpretation of Vedic rituals and texts. This system is called *Purvamimamsa.* The term is derived from the Sanskrit root "*mam,*" to think, examine, investigate. The purpose of *Mimamasa* is to inquire into the nature of Right Action, or *dharma,* which is the spiritual prerequisite of life.

The sixth and last system of philosophy is *Uttaramimamsa* or *Vedanta.* This enquires into the nature of the ultimate reality and shows how the world with its infinite variety is only an appearance, and that all things are one and the same, only appearing differently. The central theme is the philosophical teachings of the *Upanishad* concerning the nature and relationship of the three principles—God, the world, and the soul, including the relationship between the Universal soul and the individual soul. The *Vedanta* endeavors to sum up all human knowledge, presenting as truth all that is universal, and reconciling all that is different. It does not accept anything as final, but it examines, investigates, and analyzes. It has been compared to a treasure chest of spiritual insights gathered by the truth seekers of the past (Bernard, 1995).

In conclusion, the Hindu tradition seems to appear as a powerful psychological system with an emphasis on consciousness as the primary reality. In essence, this system implies that the mind and intellect alone cannot know the truth. Rather, truth must be realized as one's own being because it is one's own being. Both the philosophy and psychology of Hinduism are based on the assumption that there is enormous potential hidden in the human mind to seek answers to fundamental puzzles such as the purpose of life, the meaning of being human, the causes of suffering, the secret to a happy and peaceful life, the relationship between human energy and the universe, and so forth. These basic philosophical and psychological ideas, hidden and scattered within the ancient scriptures, offer tools for psychotherapy, education, management, and social work, and play an important role in defining the conceptual understanding of elements such as the self, knowledge and learning, diversity, and moral development within the Indian worldview.

The "self"

As philosophers discuss the concept of self and try to reconcile the many ideologies that present conflicting aspects of the self, schools and teachers turn to multicultural education to understand these various concepts. If education is to be the development of the self, then we first need to define what the self is. There are different notions of the emerging self and we can no longer assume that one kind of schooling can fit every kind of self (Gotz, 1995). Keeping this in mind, it stands to reason that in discussing education in India it is necessary to understand the concept of self as seen through the beliefs of most Indians—beliefs that are ancient but still very much alive and practiced on many different levels in people's daily lives. *Vedic* philosophy provides the base for Indian tradition, and lays down the constructs for the way life is to be lived in order to attain salvation or *moksha*.

According to this philosophy, Reality is viewed as consisting of Life and Nonlife. Life here is the self, and Nonlife is matter. Matter has three basic qualities: (1) harmony and truth, (2) energy, and (3) inertia. These three strands of matter surround different aspects of the self, and each aspect thus imagines it is a separate and independent self. Liberation takes place when each deluded aspect realizes that it has no separate identity, but is, and has always been, just the self (Gotz, 1995). The primary purpose of human life is to overcome this delusion, to find one's true self, and identify it as being separate from one's ego, thus achieving liberation or freedom from human suffering. This freedom is known as *moksha*. This can be done only when the muddiness of the material envelope made up of energy and inertia is clarified, and the self can see itself in the purity of the truth. According to the *Bhagvad Gita*, it is possible for the self to be seen in its clarity, and this realization can occur through three paths: (1) the path of meditation (*dhyanyoga*); (2) the path of knowledge (*jnyanyoga*), and (3) the path of devotion (*bhaktiyoga*). That's why knowledge, or *jnyan*, has been emphasized in Hindu philosophy for thousands of years as a way to salvation. *Upanishadic* texts underscore the development of this level of cognition in the individual: "The Self which is free from sin, free from old age, from death and grief, from hunger and thirst, which desires nothing but what it ought to desire and imagines nothing but what it ought to imagine, that it is that must be sought, this it is we must try to understand" (Chandogya Upanishads, 3, 8, 11). The scriptural emphasis on the pursuit of higher knowledge has consequently led to a high value being placed on learning and education within the average Indian household. Although rituals are still an important part of Indian philosophy, there is also the critical emphasis on higher learning and knowledge.

According to Hinduism, the four goals of life are *dharma* (morality/righteousness), *artha* (pursuit of wealth and well being), *kama* (pursuit of bodily desires), and *Moksha* (salvation). The most important goals are adherence to one's *dharma* and striving toward *moksha*. The remaining two goals, pursuit of material wealth and bodily desires, are a part of human life but are to be kept in perspective of *dharma* and *moksha*. *Dharma* is a hard-to-define concept but can be generally understood as that which maintains the order of the universe; a social cement that holds a society together. According to the *Mahabharat*, people secure mutual protection not through the state, the king, the mace, or the mace-bearer but only through *dharma*. *Dharma* is explained as being the principle and the vision of an organic society, with all the members who participate in this organic society being interdependent and their roles being complementary (Kakar, 1981). To *dharma* are ascribed such social concepts as duty, responsibility, and morality. In other words, each individual is positioned in multiple roles and must carry out the responsibilities of those roles. This would be the individual's *dharma*.

Another concept that is inherent in the way of life in India, followed however consciously or subconsciously, is the notion of *karma*. This refers to the actions of the individual human being, and the consequences of these actions. *Karma* views human actions within a cause–effect paradigm, emphasizing the careful selection of words and deeds by an individual. *Moksha*, or liberation from human suffering, cannot be attained if we do not live by our *dharma* while being mindful of our *karma*. However, the scriptures also acknowledge and recognize at the same time that *moksha* may not necessarily be attained in one lifetime and can take several lifetimes. This explains the conceptual relationship between *karma* and rebirth—a chance to better one's actions within one's *dharma* in each lifetime, and so strive toward *moksha*. Once *moksha* has been achieved there are no more rebirths. Although the *dharma* we are born into is not our choice, the chance to attain *moksha* is. It is through our conscious decisions and actions that we determine the quality of our lives. Thus *karma* is not simply a doctrine of reincarnation, nor does it imply fatalism or predestination, but rather is a promise of hope and responsibility for one's own actions (Nanavaty, 1973; Varma, 1969). Attaining *moksha* is not dependent on a third entity, but determined by the actions of our own selves. This explains the powerful urge to know the self. This Indian concept of the self is very different from the concept of the individual within the tradition of progressive education. Unlike the independent development of the individual in Western societies, the Indian self is viewed largely against a social fabric. The self is assessed within a social context, and evaluated in terms of its duties toward the group and community. An interesting distinction between the Hindu injunction "Know thy Self" (*atmanam vidhi*) and the Socratic command "Know thyself" has been made by Paine (1998):

> A westerner's identity lies geographically within the contours of his physical body and by probing there using psychological methods of examination and introspection taught since Socrates, he will uncover who he is. In India, by contrast, an individual's identity is not circumscribed by his skin's boundaries but bleeds over into the community—a community composed not only of family and neighbors but also of animals and ancestral spirits and deities as well . . . a person becomes his true self as he enters into the living stream . . . of the community life and its traditions. (pp. 101–102)

Thus, in the Indian worldview, the self is not defined by the individual alone, but, more importantly, by the individual's relationships to family and community, and through the concepts of *dharma*, *karma*, and *moksha*.

The concepts I discuss next are knowledge and learning, and why they have traditionally been given priority in Indian society and viewed as high-value commodities.

Knowledge and Learning

According to Hindu philosophy, the purpose of life is to learn about one's self, about the universe, and to understand one's own individual *dharma* in the context of the larger *Dharma*, or moral order, of the universe. This implies acquisition of knowledge in order to open up the mind and expand its cognition of the universe. In other words, this kind of knowledge acquisition is not based on rote learning but rather on comprehension. It is written in the *Upanishad*, "*sa vidya ya vimuktaye,*" which means proper knowledge is only that which frees the mind from bondage and leads it to liberation. Education was considered to be one of the most powerful channels through which an individual could attain self-knowledge, and its final goal was to uplift human character through a process of self-renewal and self-development (Nanavaty, 1973). Ancient Indian educators laid great emphasis on the acquisition of all knowledge, and believed that an individual should be accorded honor and respect based on erudition, work, age, and wealth, in that order (Achyuthan, 1974). The ideal conditions for learning were stated to be, first and foremost, the desire to learn and a love for truth; the stages in the learning process entailed receiving lessons daily, understanding them, retaining them in the memory, reflecting upon them, and exercising judgment or discrimination (Achyuthan, 1974; Vyas, R.N., 1981).

The definition of intelligence as it appears in several different Ancient Indian texts exhibits remarkable parallels in the conceptualization of this concept as understood today. Intelligence was variously defined even then. Kautilya, a prominent scholar and advisor to an emperor in the fourth century B.C., considered intelligence to be the capacity for work. In his historically famous book on policy, *Arthashastra*, Kautilya has listed the characteristics of inquiry, hearing, perception, retention in memory, reflection, deliberation, inference, and steadfast adherence to these conclusions as being qualities of the intellect. Vishnumitra, another scholar, was the author of the famous *Panchatantra* tales, fables written prior to the sixth century B.C. that later became the model for the more modern Aesop's Fables; its stories are a staple of childhood written and oral literature in India even today. In his writings, Vishnumitra defined intelligence as the power that gives us control over the world. These various conceptualizations of intelligence with its multiple dimensions are clearly related to the notion of the intellect. Additionally, it was also believed that the intellect was influenced by both hereditary and environmental factors. Schools as special places where knowledge is imparted and learning is encouraged have been in existence in India for more than 3,000 years. Forest schools existed even before 1000 B.C. as evidenced in Indian mythology and literary texts. One of the social rites of passage performed when the child was twelve years old was to mark the time when the child became a student and embarked on the educational stage of life that would be lived in the *gurukul*, the forest or community school of the *guru* or teacher (Achyuthan, 1974; Altekar, 1965).

The twentieth-century Indian philosopher and educator Krishnamurti believed that knowledge should be pursued with the intent of keeping alive the learning process. In referring to the totality of education, he considered a total human being as one who not only had a capacity to explore, examine, and understand his inner being, but also someone who was outwardly good. Thus the real goal of education must be to ensure that schools will prepare the child for both outer and inner goodness. Current education seems to focus more on helping individuals know so much about other things but they know so little about themselves. And yet, this learning about the self by traveling inward is not only a much harder journey than going to the moon, but it has to be done alone, without the guidance of books and theories (Krishnamurti, 1974). Perhaps that is why the development of concentration, focus, self-reflection, and memory were important educational ideas in Ancient India. The *Upanishad* proclaim, "through memory we know our sons, through memory our cattle." Memory was considered to be an innate attribute of the human mind. The concept of memory has also been specifically discussed and defined in various Ancient Indian texts and it has been traditionally viewed within the context of four distinct cognitive processes: *avagraha*, when a general knowledge of an external object is obtained through contact with sensory organs; *iha*, when we desire to have detailed knowledge of the object by comparing its similarities and differences with other known objects; *avaya*, when we desire to get the knowledge corroborated; and *dharana*, when the permanent impression of an object assists our knowledge (Vyas, R.N., 1981).

Discussions on the concept of diversity also appear prominently in Ancient Indian texts and I discuss that next.

Diversity

The text of *Vedanta*, composed between 500 B.C. and A.D. 200, is one of the six philosophies of Hinduism, as discussed earlier in the chapter, and is based on a contemplative approach that is a special feature of Indian life. The six different philosophies are actually six fundamental interpretations of a single reality and are complementary. The basic purpose of Indian philosophy is to eliminate suffering from human life (Varma, 1969). It seeks to do so by trying to understand the nature of that suffering and the reality behind it. Thus Hindu philosophy may be seen more as an art of life and living than as a theory about the universe. The *Vedanta* explain the unity of the universe in terms of nonduality, *advaita*, which literally means "not two." It is this very basic tenet of Hindu philosophy that drives home the fact that all diversity at a superficial level is an illusion or *maya*. But at a deeper, underlying level, all differences have a common source from which life originated. This thinking is ingrained in Hindu thought, and might explain why, in spite of multiple representations of gods and goddesses, Hindus still believe that there is only one God, one Reality, and that there is a bit of that Reality in every one of us. This belief is supported by two commonly heard Sanskrit phrases: *tat tvam asi* (that thou art) and *vasudeva kutumbakam* (the whole universe is my family).

Another point to note is that the Indian point of view is also characterized by a sense of synthesis. This explains how Indian philosophy is not pure philosophy as defined by Western standards, but synthesizes elements from religion, psychology, education, ethics, political thinking, social thought, and so forth (Vyas, R.N., 1981). This is also the primary reason why Indian dance, arts, music, pure sciences, politics, and so forth have always been closely connected to religion and philosophy, and the Indian mind has always been able to comfortably combine a scientific outlook with a spiritual outlook. Closely linked to spiritual philosophy was a scientific bent of mind, both supported by Ancient Indian scriptures. This allowed the emergence of not only philosophers and spiritual leaders in Ancient India but also scientists, astronomers, engineers, geographers, aeronautical engineers, physicians and surgeons, mathematicians, and so forth: Gargya is credited with the enumeration of 27 constellations in the *Veda*; Bharadvaja presided over the first medicinal plants symposium of the world in 700 B.C.; Attreya Punarvasu is mentioned in the *Mahabharata* as starting the Academy of Medicine; Sushruta, known globally as the father of surgery invented the technique of rhinoplasty even before the Christian era; Kanada was the first to expound the Law of Causation and the Atomic Theory in his books written in the sixth century B.C.; Medhatithi, the first to extend numerals to billions is mentioned several times in the *Rig Veda*; Aryabhata is known to have made great strides in advancing the knowledge of Algebra in the fifth century A.D.; and so forth. These are only some of the many more names that appear in texts such as the *Veda* and *Mahabharata* (Satya Prakash, 1965).

Several examples may be found in the different Indian scriptures that also imply Hinduism to be a philosophy that embraces diversity at various levels:

> The highest state (of mind) is beyond reach of thought.
> For it lies beyond all duality.
> There is only one Self in all creatures.

The One appears many, just as the moon
Appears many, reflected in water.
We see not the Self, concealed by maya;
When the veil falls, we see we are the Self.
        (*Amritabindu Upanishad*, 500 B.C.,
                pp. 243–244)

Not female, male, nor neuter is the Self.
As is the body, so is the gender.
The Self takes on a body, with desires,
Attachments, and delusions. The Self is
Born again and again in new bodies
To work out the karma of former lives.
The embodied self assumes many forms,
Heavy or light, according to its needs
For growth and the deeds of previous lives.
The evolution is a divine law.
        (*Shvestashvatara Upanishad*,
                500 B.C., p. 229)

The Lord of Love willed: "Let me be many!"
And in the depths of his meditation
He created everything that exists.
Meditating, he entered into everything.
Realizing this we see the Lord of Love
In beast and bird, in starlight and in joy,
in sex energy and in the grateful rain,
In everything the universe contains.
        (*Taittriya Upanishad*,
                500 B.C., p. 144)

In the face of various foreign attempts to suppress it, the spirit of India has been seen to continually resurge after absorbing many elements from the foreign culture. The values of tolerance and assimilation upheld by the *Upanishad* facilitate the process whereby Indian culture and society has been able to imbibe elements from other cultures. This attitude has enabled the Indian mind to broaden after coming into contact with the wider world, and absorb many fruitful ideas from other cultures, without prejudice and without abandoning any of India's traditional values (Dave, 1991; Nanavaty, 1973).

Awareness of the concepts of morality and duty is inherent in the daily lives of the Indian people to a great degree, and I address these issues in the next section.

Moral Development

As discussed above, Indian philosophy is rooted in the concept of *dharma*—that sense of duty, social obligation, morality, responsibility—which gives order to this universe. Thus moral development is not only prescribed in the ancient scriptures, but prioritized as an important goal for education as well so as to be practiced in daily life by the citizens of Indian society. This can be traced in the educational ideas in India over the course of more than 3,000 years of educational history. The greatest example of morality is presented in the *Bhagvad Gita*, the most profound rendition of Indian philosophy and such an exemplar of morality that it is the book used to swear truth by in all courts of law in India. The *Bhagvad Gita* is one of the sections in the epic *Mahabharata* that was written between 500 B.C. and 300 B.C. Before beginning to explore the position of moral development as taught in Hindu philosophy, it is necessary to clarify that moral development here is defined in terms of the attainment of harmonious and rightful conduct in the midst of diverse human desires. It includes the consideration of emotions and impulses in addition to reasoning and rational dialogue. The *Bhagvad Gita* is significant in that it provides an example of a hero facing a moral dilemma. It realizes the limits of reasoning in resolving a moral conflict, and points to the resolution lying in a dimension of human emotion that may be closer to intuition or spirituality.

In the *Bhagvad Gita*, Arjuna finds himself on the battlefield of Kurukshetra, where the mightiest battle in Indian mythology is to be fought between the Kauravs and Pandavs for succession to the throne of Hastinapur in the year 1000 B.C. Hastinapur is located near modern day New Delhi. However, Arjuna, as a result of a conflict within his own extended family, is placed in a position where he must fight his own relatives on the enemy line. As a result of his belonging to the noble warrior caste, the *Kshatriyas*, Arjuna's duty is to fight. Yet at the same time he recognizes his duty toward his family, and questions how he can fight and kill them. Katz (1989) explains that the dilemma arises due to the two conflicting, yet equally compelling, social duties: his sense of duty and responsibility toward the elders in his family whom he must oppose, and the social obligation as a warrior (defined by the term *kshatriyadharma*) as part of his social duty toward his society as a whole (defined by the more universal *sanatandharma*). Both cannot be honored simultaneously, and Arjuna must decide to which *dharma* he is morally expected to adhere.

Zigler (1994) provides an assessment of this situation in terms of Western theories of psychology, drawing upon both Kohlberg's and Gilligan's perspectives on moral development. Lawrence Kohlberg's concept of morality draws from the traditional male expressions of moral virtue and, in Piagetian tradition, is marked by its emphasis on individual rationality, autonomy, independence, and self-interest. Carol Gilligan, on the other hand, argues that the Kohlbergian model is inapplicable to the way women conceive of their moral dilemmas. Women's approach is marked by an ethic of care and nonviolence and reflects a sense of connection and responsibility for others. In this case, Arjuna is clearly overwhelmed by his conflicting responsibilities arising from his sense of connectedness not only with his relatives but also with his society as a whole. A decision either way, to fight the battle or not fight it, is equally painful for him. It is in this no-win situation that Arjuna laments, "we do not know which is better for us: that we should conquer them or they should conquer us . . . if we killed them we should not wish to live" (*Bhagvad Gita*, 2–6). There is a distinct issue of balance here, between individual autonomy and our interdependence with the social environment.

The state of the mind is critical in the resolution of such a moral dilemma. Arjuna is in a state of emotional and physiological disequilibrium as he cries out: "my limbs grow weak; my mouth is dry, my body shakes, and my hair is standing on end. My skin burns, and the bow *Gandiva* has slipped from my hand. I am unable to stand; my mind seems to be whirling" (*Bhagvad Gita*, 1–28, 29, 30). This disequilibrium is brought on by his sense of fear, not for his own life but his concern for the downfall of morality, of the standards that hold society together. At this stage, it is his mentor Lord Krishna who as his charioteer offers him counsel. His advice is to first regain a state of equanimity—a balanced and disciplined state of mind and body. This balanced state of mind may be achieved through *dhyana yoga* or the yoga of meditation. According to Krishna an individual who is established in *yoga* does not suffer from the fight or flight alarm responses, and consequently does not incur sin by making the wrong decision. At this point, Arjuna in a more balanced and spiritual frame of mind understands that it is his *dharma* as a warrior that he must adhere to. He must fight this battle in order to carry out the larger *dharma* of the universe by preventing the disintegration of morality in society. A fundamental spiritual truth is realized, which transforms Arjuna, who then resolves his moral conflict and returns confidently to the battleground without actually going through a process of reasoning based on rationality and logic. Arjuna is seen to have realized his Self through a three-pronged process that consists of action, knowledge, and devotion (Nanavaty, 1973).

In her analysis of Arjuna's character in the *Mahabharata*, Katz (1989) appreciates the significance of studying Arjuna because he is revealed to be not only the ideal hero (as an upholder and protector of the right in society), but also a human being, the everyman, who goes through an emotional breakdown in attempting to resolve a moral dilemma, as well as a devotee with a spiritual side. The reassuring conclusion is that a person, a human being, can possess the qualities of a hero and a devotee simultaneously. If a spiritual discipline illustrates physical well-being and moral growth outside the context of religious dogma, then there is no reason why these disciplines should not be included in education, and it would, in fact, be irresponsible not to do so (Zigler, 1994).

The discussion on the texts and philosophical concepts presented above was done with the purpose of understanding the system of education presented in chapter 3 within the sociocultural context of Indian society. However, an understanding of the sociocultural complexity of Indian society is incomplete without a discussion of the multiple layers of diversity that exist within it. Thus, a section on diversity is included here to present to the reader the massive range of cultural variation that can be seen and experienced in India.

The Diverse Fabric of Indian Society

Over the last several thousand years, India has existed as a nation of vast cultural and religious diversity, and this multiethnic, multireligious, multilingual country populated by more than one billion people has been called one of the world's great wonders (Friedman, 2002). A commonly heard cliché in India is "unity in diversity," and this is a phrase that is not restricted to the identity of modern India only. Written documents describe foreign travelers in the past who have observed and recorded this fact about Indian society: travelers and historians such as Megasthenes in 315 B.C., Fa-Hien in A.D. 404–411, Huan Tsang in A.D. 630–644, Alberuni in A.D. 1030, Marco Polo in A.D. 1288, and Ibn Batua in A.D. 1325 (Dube, 1990/2000).

India's diversity is mainly the result of the different racial, religious, linguistic, and ethnic backgrounds of her people. Six major racial elements in the population of India have been listed by Dube (1999/2000): Negrito, Proto-Australoid, Mongoloid, Mediterranean, Western Bracycephals, and Nordic. The first three groups are the oldest residents of the Indian subcontinent and are now located in small pockets: Negrito groups in southern parts of India, on the western coast, and on the Andaman and Nicobar Islands in the Indian Ocean; the Proto-Australoid group in the tribes of central India; the Mongoloid groups in the Himalayan and the northeastern regions of India. The next three racial groups were said to have arrived in India about two thousand to three thousand years ago. People from the Mediterranean regions were associated with the Dravidian cultures of

Ancient India: the western Brachycephal characteristics are seen in population groups in northern and western India, and the Nordics included the Indo-Aryans who arrived in India after the Indus Valley Civilization. The Indus Valley Civilization is said to have flourished around 2300 B.C. and its people were already a multiethnic population as suggested by excavations and skeletal remains (Dube, 1990/2000). Much later came waves of immigrants such as Greeks, Scythians, Parthians, Kushans, and Huns who settled down in India and were absorbed into the Indian social system.

In addition, India has nurtured populations from different religious backgrounds for centuries. Dube (1990/2000) names eight major religious communities in India with each religion subdivided into religious doctrines and sects. Hindus constitute roughly 82 percent of India's population and can be divided broadly into *Shaivite* (worshippers of *Shiva*), *Vaishnava* (worshippers of *Vishnu*), *Shakta* (worshippers of the Goddess in her various forms), and *Smarta* (worshippers of all three), with each group having further subdivisions and sects. The Indian Muslims constitute about 12 percent of the population and are divided broadly into *Sunni* and *Shia* communities. Indian Christians form about 2.6 percent of the total population in India and are further divided into Roman Catholics and Protestants with many denominational churches. Other religious groups in India are Sikhs (2 percent), Buddhists (0.7 percent), Jains (0.4 percent), Zoroastrians (0.3 percent) or the Parsi people who came to India from Persia in the eighth century A.D., and Jews (0.1 percent), members of the Jewish faith, which was established in India over 1,000 ago. Although 0.1 percent might seem a very low number, in the context of the Indian population it would still amount to more than 1 million Jewish people in India itself. The legal system in India supports separate personal law codes for Hindus, Muslims, and Christians, and this helps to protect the rights of minority groups.

Language is another source of great diversity in India. The Indian Constitution lists 22 official languages, which is still a bare minimum estimate. Dube (1990/2000) provides an example of how this arbitrary official number runs into practical difficulties in the awarding of annual literary awards by the Sahitya Academy (a national academy that honors literary works), which has to recognize nonofficial languages such as *Dogri, Manipuri, Konkani, Maithili, Nepali,* and *Rajasthani* as separate literary languages. There are 227 mother tongues recognized in India, and approximately 1,600 dialects. The national language of India is Hindi, and even this has several dialects such as *Awadhi, Bagheli, Bhojpuri, Brij, Bundeli, Chhattisgarhi, Hadoti, Magahi, Malwi, Nimari, Pahari, Rajasthani,* among others.

The Indian caste system is an extremely complex phenomenon with innumerable classification systems in different parts of the country. An Indian has several identification tags and identity is based on religion, place of residence, family, a caste with an ascribed status, member of a group as having descended from a common ancestor, and a lineage that goes back several generations. Traditionally, ascribed status led to the recognition of four classes; *Brahmans* the priests, *Kshatriyas* the warriors, *Vaishyas* the merchants and traders, and *Shudras* the artisans. There was a fifth level, people who engaged in "unclean" occupations such as sweeping, cleaning toilets, and the like, and this group was called the Untouchables. Although "untouchability" has been legally abolished, its practice continues in either unconscious or conscious ways in many parts of the country. This class, now known as the Scheduled Caste, constitutes about 15 percent of India's population and is protected under several affirmative action rulings. Another class is called the Scheduled Tribe representing about 7 percent of India's population and includes members of indigenous tribes. But although the social system seems clearly organized at a theoretical level, the reality is far from simple and there is no uniform hierarchical system: one type of ranked status in one region may be viewed very differently in another region.

Another source of diversity is regional cultures. The people in a certain region may have different racial, religious, ethnic, and linguistic features but a common culture based on thousands of years of historical experiences, devotional poetry, art and architecture, ecology, and the environment of the region. An early text from the south of India, *Tholkappiyam*, written in A.D. 200, profiles how society was organized in the hill areas, forests, cultivated plains, coastal areas, and desert areas (Dube, 1990/2000). Despite this tremendous diversity, Hinduism is the dominant religion in India and has influenced the ethos of Indian society over the course of several centuries. However, it is quite clear that the culture has also incorporated elements from other religions and traditions.

The Postcolonial Dilemma: Indian or Hindu/Muslim/Sikh/Christian . . .

It is important to emphasize that Hinduism referred to the way of life of a certain group of people living in a specific geographical region rather than to a religion, and Keay (2000) provides a detailed account of the origin of the word "*Hindu.*" The Ancient Indian civilizations of Mohenjo Daro and Harappa flourished in the Indus valley of northwestern India formed by seven rivers consisting of the river Indus and its tributaries. The Sanskrit word for seven is "*sapt*" and for river is "*sindhu*" and thus the region came to be called *sapta-sindhu*, or the land of the seven rivers. With time this area came to be known as *sindh*. Much later the initial "s" was rendered by the

Persians as an aspirate and became "*hind*," and the people living there came to be known as "*hindus.*" Even later, as the word found its way into Greek, the aspirate was dropped altogether and the word became "*ind*" and the name India gradually came into being.

The *Veda* and other scriptures such as *Upanishad* and *Bhagvad Gita* offer detailed recommendations for a good way of life and living, indicating that a life lived within one's *dharma* and mindful of one's *karma* is the way to achieve salvation. Thus, it is possible to be a good Hindu even if one is not religious and never visits the temple. The religio-philosophical scriptures such as the *Veda*, *Upanishad*, and *Bhagvad Gita* are not remote from popular life in India, and are seen to permeate various aspects of people's lives. Deeper truths of life are seen to prevail among the people of India in the form of myths and stories from the *Purana*, *Ramayana*, and *Mahabharata* (Dave, 1991). It is a fact that *dharma* and *karma* are technically Hindu terms, and the philosophical and psychological ideas that I have introduced in this chapter have a tremendous emotional appeal to Hindus. However, in Indian society many of these concepts are internalized by not only Hindus but by non-Hindus as well; hence, over time, it has become a universal way of life. Even when Buddhism, Jainism and Sikhism were founded as religions separate from Hinduism due to disagreement with certain Hindu concepts, the actual practice of some core Hindu concepts could not be shed. Similarly, the deep-seated beliefs of the Hindu tradition as a way of life could not be discarded by many of those who converted to Islam and Christianity (Dube, 1990/2000). In other words, it can be said that key aspects of the Vedic/Hindu philosophy have been integrated into the daily living of people in India, transcending religious beliefs. However, having said this, I also want to say that Indian culture is not only a Hindu culture but reflects ideas and practices from all the different subcultures and religions that have accumulated over the centuries and that are found in this mix. Some of the religions found in India are themselves based on an eclectic mix of ideas. For example, both Buddhism and Jainism are variations of Hinduism. Sikhism draws on both Hindu and Islamic ideas. The Urdu language that was used widely in the past, and is used extensively in Muslim dominated areas of India even now, evolved as a combination of Hindi and Farsi during the Mughal rule. The influence of Muslim practices is seen widely in much of Northern India where the Muslim strongholds used to be. Children's literature includes stories from the past and the present featuring prominent characters who may be Hindu, Muslim, or Christian. But it is also reasonable to assume that a nation's general culture reflects the beliefs and behaviors of the dominant groups. Since 80 percent of India's population is Hindu, Hindu beliefs and practices are seen to dominate Indian culture. However, there is a high degree of complexity in the conceptual understanding of the "Hindu-Indian" dynamic, which I will try and illustrate through several examples.

A common Vedic custom that is practiced by many Hindus is the light-ing of the "*diya*" (a clay lamp) at the altar before sitting down to partake of the evening meal. A past acquaintance who used to be a hospital nurse and was Christian by faith had adopted and adapted that Vedic custom. She would not eat the evening meal until she had lit a candle in her neighbor-hood church on the way home. Her family had also begun to incorporate this ritual into their routines. This certainly may not be a practice that a Christian family in another country might engage in.

Another example is the practice of *yoga*. *Yoga* is one of the six *Darshana*, or schools of Hindu Philosophy, as has been described earlier in this chapter. But the practice of *yoga* as a way of life and the recognition that it offers physical, mental, and spiritual benefits has transcended geographical boundaries and religions. Today, *yoga* is a popular health practice viewed by non-Indians everywhere as an Indian practice and not a Hindu one.

Another example is with reference to the well-known phenomenon of Indian hospitality. The importance that has been and continues to be given to a guest in India draws from an age-old Hindu belief. There is an ancient Sanskrit saying "*atitih devo bhavah*" (your guest is like your God and deserves the respect and importance one would accord to God). Although this is originally a Hindu saying in Sanskrit, the concept is practiced widely across religious, linguistic, and socioeconomic groups in India and is con-sidered by all Indians as an Indian custom, not a Hindu custom specifically. On one of my recent visits to India, I came across a full-page color adver-tisement, in a prominent newspaper, that had been sponsored by a success-ful Indian car company. The advertisement declared in large bold type "*Atithi Devo Bhava: we found our mantra for customer delight in every Indian home*" (November 26, 2004). I thought of it as a good illustration of the sentiments that I have been trying to describe. There are critics who view the use of Hindu images by the media as an attempt to colonize the Indian masses and impose the hegemonic views of the dominant Hindu culture on all Indians. But there are also an equal number of critics who are disdainful of the Westernization of the Indian masses through Macdonaldization and Disneyfication. Certainly the question arises as to which images would be more appropriate and less colonizing. In many ways, the Indian media only seems to mirror the culturally hybridized mix of ideas that already exists in the minds and lives of people in urban Indian society.

The multilingual phenomenon in India presents another interesting aspect of "Indian-ness" transcending different religions. In India, a country with 22 official languages and almost 1,600 dialects, languages are not spe-cific to a particular religion. Hindus, Muslims, and Christians in India speak the language of the state they have roots in. For example, Christians in Kerala may have Malayalam as their mother tongue whereas Christians in Tamil Nadu may have Tamil as their mother tongue. The official language in Kolkata (formerly known as Calcutta) is Bengali and probably is the mother tongue of all those who live in Kolkata be they Hindu or Muslims or Christians. Many languages in India are derivatives of archaic Sanskrit, the language of the ancient Vedic texts. And since language plays a vital role in the sociocultural construction of knowledge and in the cognition of experiences, many Vedic concepts have become an Indian way of life: things come to be considered as just Indian by the multitudes of people who actually live and speak those realities in their daily lives.

The following excerpt is from a book recently written by A.P.J Abdul Kalam, the current president of India:

> "When you speak, speak the truth; perform when you promise . . . What actions are most excellent? To gladden the heart of a human being, to feed the hungry, to help the afflicted, to lighten the sorrow of the sorrowful and to remove the wrongs of the injured . . ." These are the sayings of Prophet Mohammad. My friend who told me this is a greatgrandson of a Deekshidar of Tamil Nadu and a Ganapaathigal (vedic scholar) . . . Such an outlook is possible only in our country. Let us remember the Rig Veda: Aano bhadrah kratavo yenthu vishwathaha: That is "Let noble thoughts come to us from every side". (Kalam, 2002, p. 95)

This quote is from his book *Ignited Minds* and is an example of how ideas are exchanged amongst Indians with different religious backgrounds. The Vedic scholar is comfortable quoting from the *Kuran*, as is the author, an Indian Muslim, recalling words from the *Rig Veda*. Hundreds of years ago, Indian thought was divided into philosophy, and religion by colonizers to better fit the Greek definition of philosophy, which has become the standard approach to philosophy in the West (Dhillon, 1997; Ozmon & Craver, 1995). Today, one must ask the question whether a desire to differentiate what in "Indian" is Hindu, Muslim, Sikh, Christian, and so forth is felt as intensely by those who actually live the coexisting multiple realities of life on a daily basis within urban Indian society as it is felt by those outside it.

Thus far in this chapter I have tried to present the basic and important ideas underlying Hindu philosophy and psychology. Although I have labeled this discourse as being Hindu since it is based on Hindu and Vedic texts, I have also attempted to illustrate, by establishing the connections among the philosophy, the psyche, and the corresponding goals of education within Indian culture, that core concepts of this Hindu discourse have permeated the ethos of Indian society and have developed into an Indian way of life. This assumption will be strengthened perhaps as the reader begins to hear the voices of the participants in the research unfolding in the ensuing chapters.

Chapter 3

Educational Systems in India: Past and Present

India is the world's largest democracy with about 1.08 billion people and a civilization that spans more than 5,000 years. At the time of independence in 1947, after the almost 300-year-long British rule, India was divided into two separate countries, India and Pakistan. This separation occurred due to differences between the Hindu and Muslim political leaders at that time, and the partitioning of India by the departing British administrators is a period of great sadness, violence, and bloodshed in the history of the subcontinent. Pakistan emerged as an independent nation consisting of West Pakistan (on the western border of India) and East Pakistan (on the eastern side bordering the Indian state of West Bengal). In 1971, further conflicts resulted in another geographical and political division, with East Pakistan claiming independence and emerging as yet another separate country now known as Bangladesh.

The concept of education in India today can only be understood clearly when one views the development of education within the context of India's history, the relationship between education and India's cultural and spiritual philosophy, and the impact on education of key social–political–cultural influences from foreign invasions. A recorded history of formal and informal education in India dates back to between 3,000 and 4,000 years. Education in Ancient India was based directly on Vedic philosophical verses and scriptures compiled in an archaic form of Sanskrit as early as 2000 B.C. In 1835, a British survey by William Adam reported the existence of an extensive system of popular indigenous education throughout India even before British initiatives were implemented (Keay, 1918/1980). According to the report, in the 1830s there existed more than 100,000 schools in the Indian states of Bengal and Bihar alone, resulting in the fact that almost every village had its own school (di Bona, 1983; Goyal, 2003). This extensive system of education, even in the early nineteenth century, was relatively free from caste and religious discrimination and a large number of villages had schools. Written records indicate that even from 250 B.C. to A.D. 250 free education for all except for unskilled workers was provided throughout India (NCERT, 2001; Saini, 2000).

Over the centuries, India came under a series of foreign conquests and rules, most significantly, the Islamic invasions from Asia Minor and Persia, and the Christian colonizers from Europe. Subsequently, under these various cultural, religious, and political influences several systems of multifaceted education were implemented. During the approximately 3,000–4,000 years of formal and informal education in India, a historical evolution of education may be traced through the distinct and sometimes overlapping influences of the Hindu/Vedic, Buddhist, Muslim, British, and the post-Independence periods (Basham, 1998; Keay, 1918; Paranjoti, 1969; Vyas, K.C., 1981). In this chapter, I have briefly highlighted aspects of education such as philosophy, curriculum, and significant policies as they appeared in the various historical eras to provide a clearer picture of the process of evolution that education underwent in India. These details facilitate the understanding of the extensive influence that various cultural and spiritual philosophies had on the system of education before, during, and after the British colonization of India. This knowledge also facilitates a better understanding of the layers of complexities in the current fabric of education in India, and positions more accurately a study of the same within a postcolonial paradigm.

The Influence of Ancient Hindu Texts and Scriptures on Education

Hindu texts and scriptures of Ancient India formed the basis of the very first recorded educational system in India that dominated from approximately 2000 B.C. to A.D 700. During that time education was based primarily on the teachings of the *Veda* and *Upanishad*, and was characterized by deep philosophical reflections, spiritual and secular learning, and intellectual explorations. The main features of some of the Ancient Indian texts and scriptures have already been described in detail in chapter 2. Education was available to most during the early Vedic stage but became more discriminatory as the caste system began to evolve. The Vedic texts from about 2000 B.C. are the earliest known writings on existential philosophy; Kautilya's writings from approximately the third or fourth centuries B.C. provide detailed studies of politics and education; Manu's writings, dated to the early years of the Christian era, are considered to be the first law books in India; and the great Indian epics, the *Mahabharata* and the *Ramayana*, include detailed references to social norms, teachings, and education of the times. Specific details in terms of curricular content areas as recommended by Manu and Kautilya may be found in the works of Achyuthan (1974) and Keay (1918/1980), among others. The famous Indian scholar Panini wrote the Sanskrit grammar in the fourth century B.C., and it is considered the earliest known form of organized grammar. It was the discovery of Panini's grammar that led to the study of phonetics in the West. Besides the study of the *Veda*, several other subjects were included in a typical curriculum in Ancient India. The *Rig Veda*, the first book of the *Veda* and earliest known writings, emphasized that knowledge could be gained only after the study of several subjects, and Vyas R.N. (1981) has provided a comprehensive list of content areas in a typical Vedic curriculum in 1500 B.C. Some of the subjects included:

- *Shiksha*: the teaching of proper pronunciation and recitation of the *Veda*
- *Kalpa*: the rules of sacrifice
- *Vyakarana*: the study of grammar and derivation
- *Chandas*: knowledge of composition, versification and meter
- *Jyotisha*: the understanding of the secrets of nature in a scientific manner
- Reasoning, including logic
- The sciences: medicine, astronomy, physics, chemistry, and so on
- Metaphysics: whereby the concept of Reality was seen as being infinite: "*purusa eva idam sarvam yat bhutam yat ca bhavyam*" (Reality is all that is, has been or will be).

In addition to these content areas, Vedic education also taught people the skills to become efficient in their occupations as carpenters, artisans, physicians, farmers, poets, and so forth. There were no class or gender distinctions at that time, and according to the third Vedic book, *Yajur Veda*, Vedic knowledge could be imparted to anyone. There are several instances in Vedic literature that describe the participation of women in higher education, and Apala, Paulomi, Gargi, Urvashi, and Devyani are some among many others named. This is not surprising as the Hindu religion recognizes the equal importance of the male and female energies in gods and humans, and the goddess figure occupies a powerful position in Hinduism. The progressive nature of Vedic education as education for all, disclaimed social and economic barriers and implied a rare gender-equality in society. Music was an important aspect of Vedic education. Many melodies were discovered and it is said that the seven *svara* or notes on which Indian music is based were known in Vedic times. Medical science was developed and systematically taught to students and there is mention in the *Veda* of herbal medicines for conditions or diseases such as fever, leprosy, jaundice, dropsy, cough, baldness, snakebite, and mania, among others. However, over subsequent centuries the business of higher education became more concerned with the requirements of priesthood as the rituals of sacrifice began to gain increasing prominence in Vedic society.

The *Upanishad* are scriptures that were written after the *Veda* in about 800 B.C. These scriptures also have strongly influenced Indian education. Steeped in the spirit of inquiry, of mental adventure, and a passion for finding out the truth about things without tolerance for any dogma standing in the way, the *Upanishad* encouraged the study of the external world as an aspect of the inner reality; interest in magic and supernatural elements was discouraged. There was a continuous attempt to harmonize social and spiritual activities. The learning process was permeated with a spirit of exploration and both teachers and students were considered to be cotravelers on a quest toward truth through discussion, discourse, and reflective dialogue. The emphasis of the curriculum lay on self-realization and knowledge of the individual self and the absolute self. Vyas R.N. (1981) describes at least 24 various subjects, the study of which was included in a curriculum within an educational system defined by the *Upanishad*. Some of the content areas in a typical 800 B.C. *Upanishadic* curriculum include the following:

- *Anushasan*: The study of systems such as phonetics, grammar, metrics, astronomy, knowledge of rituals
- *Vidya*: The study of the six philosophical systems, *nyaya, yoga, vaisheshika, sankhya, purvamimamsa, uttaramimamsa*
- *Vakovakyam*: The study of theological discourse, art of argumentation
- *Itihaas Durana*: The study of figures from legends and myths
- *Akhyana/Anvakhyana*: The study of stories and postnarratives
- *Anuvakhyana*: The study of the explanations of various *mantras*
- *Brahamana*: The study of books on religious explanations
- *Kshatyavidya*: The study of the science of wielding weaponry
- *Rashi*: The study of numbers and arithmetic
- *Nakshatra vidya*: The study of the science of astronomy
- *Bhuta vidya*: The study of the science of life forms
- *Upanishad*: The study of knowledge about the supreme reality
- *Ved Veda*: The study of the grammar of the ancient *Vedic* texts
- *Dev vidya*: The study of the science of the worship of gods
- *Brahma vidya*: The study of the different branches of the *Veda*
- *Deya jana vidya*: The study of arts such as the making of perfume, dyeing, dancing, singing, playing musical instrument.

As is quite apparent from the above, education based on Vedic and Upanishadic texts was not restricted to only religious study but in fact included a curriculum that supported the study of many different content areas and ideas. It was firmly believed at that time that a study of a combination of subjects on practical utility as well as on spiritual upliftment would lead to the development of a balanced personality. The Upanishadic passion for finding out the truth about things utilizes a scientific approach, that is, the science of reasoning and questioning, as can be inferred from the frequent use of terms such as *prasnin* (questioner), *abhi-prasnin* (cross-questioner), *prasna-vivaka* (answerer). The curricular content was divided into two categories—*preyas* (leading to worldly pleasure and utility), and *sreyas* (leading to salvation from worldly suffering).

It was about this time in history that the concept of social classes began to emerge under the then existing social influences, but they were recognized as being different limbs of society. Education was largely open to students from the *Brahman* (priestly), *kshatriya* (warrior), and *vaishya* (merchant and trading) castes. The study of the *Veda* was common to all of the above, and thereafter students could be trained in occupation-related areas of study. Thus, there was both spiritual and secular education available to most except members of the *sudra* (workers and artisans) caste, who, although they could train in menial jobs, were denied religious and spiritual learning. Basing his conclusions on ancient scriptures and texts, Altekar (1965) summarized the general aims of education in Ancient India as being an infusion of a spirit of piety; character formation; personality development; inculcation of civic and social duties; promotion of social efficiency; and preservation of the national culture.

In the *Upanishadic* period, Benares (modern day Varanasi) became a major center of learning and reference is made to Ajatshatru, one of the kings of Benares who features as a philosopher in the *Upanishads* (Altekar, 1965). In the seventh century B.C., Benares was probably the most famous center of education in eastern India, and continued to be a prominent center with the rise of Buddhism and Islam in India. The Benares Hindu University is a prominent center of learning even today and is considered by many to be one of the oldest existing universities in India. Schools that impart traditional Vedic education do exist even to this day in Indian society but not as part of the mainstream educational system. These schools can be found in places that are usually associated with temples and are attended by students training to become Hindu priests. In addition to the temple schools, there are many major universities and centers that offer courses in Vedic education and the study of Sanskrit texts. Recent interest in Vedic education has also led to government and nongovernment-sponsored initiatives for starting schools based on the concept of the ancient *gurukuls* and other Vedic schools.

The Influence of Buddhism on Education

Buddhist influence on the educational system in India occurred approximately between 600 B.C. and A.D. 700. Buddhism as a religion developed as an offshoot of Hinduism. Although Buddhist philosophy is embedded in ancient Hindu philosophical ideas such as *dharma* and *karma*, it rejected the rigidity and dogma that had begun to predominate Hinduism by that time in history. Thus the major change in Buddhism was the rejection of rituals, ceremonies, and caste discrimination, and education was made available to any person, including women, who desired to learn. Learning continued to be based on the *Veda*, *Upanishad*, as well as Buddhist scriptures such as *Dhammapada*. There are written records of scholars from around the world including China and Greece, who came to study at universities such as *Takshashila*, *Nalanda*, and *Vikramashila*, which were established in India around that time. Such was the reputation of these powerful centers of higher learning that there was a long waiting list of scholars from all around the world with the admission rate being only 20 percent (Vyas, R.N., 1981). Written records by Chinese scholar I-Tsing regarding the curriculum of studies at *Nalanda* indicate that the study of grammar received great attention and grammar was the foundation of all other studies (Keay, 1918/1980). He describes five bodies of knowledge or *vidya*: *sabdavidya* (grammar and lexicography); *silpasthanavidya* (arts); *chikit-savidya* (medicine); *hetuvidya* (logic); and *adhyatmavidya* (science of the universal soul, or philosophy). The main course was founded on an elaborate study of Sanskrit grammar, which led onto logic and finally to metaphysics and philosophy.

Buddhism was founded by a Hindu prince, Siddhartha, who later came to be known as Gautam Buddha. As a religion, it flourished for a while in most of India, especially during the reign of Emperor Ashoka and the Maurya dynasty. Ashoka is credited with its spread beyond India into other parts of Asia and as far east as Japan. Although Buddhism continued as a religion in India for 1,500 years, it did not find lasting success in India itself where Hinduism was, and has been, the dominant force; consequently, Buddhist schools gradually diminished in number. Altekar (1965) refers to *Life and Travels*, the work of Chinese scholar Huan Tsang who studied at some of the Buddhist universities named above, and mentions that even though Buddhism was on the decline in India by the seventh century A.D., many monasteries continued to be flourishing centers of Buddhist learning. Today, Buddhist education is imparted in monasteries in the northeastern regions of India, which are closest to predominantly Buddhist countries such as Tibet, Bhutan, and Myanmar, as well as in Dharamsala, which is the home of the Dalai Lama, who is in exile in India from Tibet. Buddhism is still practiced actively and extensively in many other Asian countries such as Sri Lanka, Thailand, Cambodia, Vietnam, and Japan, among others.

The Influence of Islamic Rules on Education

Islamic influences on the educational system in India prevailed in the years between A.D. 1000 and the 1700s. This period marks the start of several Islamic invasions into India from Persia and Asia Minor, and the establishment of monarchies such as those of Mahmud of Ghazni in A.D. 1001, Muhammad Ghauri in A.D. 1175, Qutub-ud-din Aibak who made Delhi the seat of Islamic power in A.D. 1193, Bhaktiyar Khilji who destroyed the famed Nalanda University in A.D. 1200, Raziya Sultana who became the first woman ruler of Delhi in A.D. 1236, and Muhammad bin Tughlaq who came to power in A.D. 1323. This period culminated in the establishment of the Mughal dynasty by Babar in A.D. 1526, which remained in power for the next 300 years. During this period, the system of education in India was marked by the development of Islamic schools that were closely associated with mosques. Education in these schools were intimately informed by the Islamic religion. Gender segregation and discrimination is a distinct characteristic that occurred in Indian society and education under the influence of Islam. Women could not be seen or heard in public, and with the implementation of the *purdah* system, women's lives in Muslim-dominated regions of India became even more sheltered and homebound with decreasing rights to education in general, as was the case with members of certain social castes. Educational philosophy advocated conformity and discouraged the critical thinking or speculation that had been valued in Hindu and Buddhist educational philosophies.

The development of a healthy encounter between Hindu and Muslim cultures was possible in the reigns of those kings and emperors who were keen to promote the meeting of these cultures. Some prominent names in the spread of education are the early Muslim monarchs of the Khalji and Tughlak dynasties in the twelfth and fourteenth centuries; Raziya Sultana in the early thirteenth century; and emperors of the Mughal dynasty from the sixteenth and eighteenth centuries namely Humayun, Akbar, and Jehangir (Keay, 1918/1980; Paranjoti, 1969). In fact, during the reign of these emperors, education was considered to be the birthright of every citizen and indigenous education was greatly promoted. A college was in existence in Delhi when Raziya Sultana came to power in the early thirteenth century. Raziya herself was a patron of learning and an educated woman, and under her patronage, as well as that of Nasir-ud-din (1246–1256) and Balban (1266–1287), literary societies flourished in Delhi; by the end of the thirteenth century Delhi had become a great center of learning and continued to be so (Keay, 1918/1980). High standards of education under the rules of other Muslim rulers also led to the establishment of many village schools and colleges. Some prominent centers of learning were those at Firuzabad, Badaun, Agra, Jaunpur, Bijapur, Golkonda, Malwa, Multan, Bengal, among others.

Humayun, who ruled from 1530 to 1556, was the second Mughal emperor and an accomplished scholar who built a college and library in Delhi. He actually met his death by falling off from the balcony of his library. Humanyun's son Akbar (1556–1605) is known to have been instrumental in the spread of education and learning not only of Islamic literature and arts but also of Hindu texts, arts, and music. During Akbar's reign, Hindu literature, such as *Ramayana, Mahabharata*, and the *Atharva Veda*, were translated into Persian. He even tried to initiate a new but short-lived religion called *Din-Ilahi*, which eclectically blended the spiritual aspects of Hinduism and Islam. There were other Muslim kings and rulers under whom Muslim influences prevailed, including the great importance given to Persian poetry, music, dance, and the arts; Persian, or Farsi, being made the language of the court in India; and a rise in the teaching of Arabic and Persian languages. In schools that encouraged both Hindu and Muslim students, a new language known as Urdu emerged from the mingling of Hindi and Farsi. The study of the *Koran*, the sacred book of the Muslims, entered the life and culture of the country. Education was largely defined by the learning of the three Rs, religious instruction, and other subjects such as literature, languages, grammar, logic, metaphysics, science, and politics (Paranjoti, 1969). Islamic education in India today is imparted, but restricted to, within *maktabs* and *madrasahs*, which are religious schools and colleges run by mosques, and in centers of Islamic studies. After Hinduism, Islam is the second-largest religious group in India with about 12 percent of the Indian population being Muslim.

The Impact of British Colonialism on Education

From about A.D. 1600 onward, the presence of French, Portugese, Dutch, and British traders and subsequent trading companies in India led to the start of European and Christian influences on the educational system. This stage saw the emergence of missionary schools in India with their predominantly Roman Catholic beliefs and Christianity's focus on individual salvation. Later, as the British gained supremacy and control over all of India, Protestant forms of Christianity as well as education based on the scientific methods of modern Europe were implemented. Under the influence of Britain's imperial rule, Western values permeated the intelligentsia and literati of India (Saini, 2000). The turning point of the British influence on education in colonial India was the adoption of English as the language of instruction in all the schools supported by the British administration. Sir Warren Hastings, Lord Macaulay, and almost every viceroy of India were instrumental in shaping education in India under the British rule. A detailed and comprehensive account of the history of English education in India from 1781 to 1893 appears in a historical document by Mahmood (1895). This includes a rare chronological account of the extracts from Parliamentary papers, official reports, dispatches, minutes, and resolutions of the government. Highlighted below are some of the more important educational initiatives in nineteenth-century British India:

- 1813: Sanskrit learning was to be encouraged for its excellent systems of ethics and codes of law.
- 1815: Missionary movements for starting colleges of education were begun.
- 1829: English was adopted as the language of official business in colonial India.
- 1835: The debate between comparative benefits of English and Oriental learning resulted in Lord Macaulay's Minute, which proclaimed English to be the language of instruction in schools in India.
- 1844: Lord Hardinge supported the employment of those native Indians who were successful in the system of education administered by the British.
- 1854: Governor General Lord William Bentinck declared religious neutrality in education in schools. This was received with much disapproval by the missionaries and was followed by the gradual conversion of Hindu students in Missionary schools.
- 1857: The Universities of Calcutta, Madras, and Bombay were founded, modeled after the University of London. Sanskrit, Arabic, and Persian languages were included among the subjects offered, but only when consistent with religious neutrality.
- 1882: The Indian Education Commission was established to report on the subject of education. Members included both European and Native Indian representatives.
- 1889: The Indian Education Commission proposed that schools include textbooks for moral education, and yet stay secular in terms of religious neutrality.

- 1892: By this time a total of 144 colleges had been established by the administration in British India. Of them there were 108 liberal arts colleges, 28 law colleges, 4 medical colleges, and 4 engineering colleges.

Another legacy of the British colonial era is the textbook-centered pedagogy that began, and continues, to dominate education in India. In the mid-nineteenth century, the colonial administrators implemented a new bureaucratic format for the educational system in India. The new system would be governed by a bureaucracy tightly controlling all aspects of schooling, the purposes of which were (1) to acculturate Indian children and youth in European attitudes and prepare them to work at the lower and middle levels of colonial administrative services; (2) to insist on English as the medium of instruction; (3) to have indigenous schools conform to the syllabus and textbooks prescribed by the colonial government if they wanted to seek government aid; and (4) to use centralized examinations as assessment tools to determine eligibility of students for promotions (Kumar, 1992). Employment in the public services and private firms became conditional on a command and mastery of the English language and literature, and British culture and etiquette. Thus, the rote memorization and examination-based system of education became largely a means whereby a class of Indian clerks and low-level bureaucrats were produced to fill key administrative positions for British officers (Saini, 2000). Much of this educational approach is still in practice and shapes the overall system of schooling even today.

In this chapter, I have presented only a brief timeline of select educational reforms in India, but a discussion on the implications and lasting consequences of such initiatives has been undertaken in later chapters.

Pre- and Postindependence Efforts

The next stage of educational change in India appeared after the end of British colonial rule when India gained independence in 1947. One movement that occurred immediately before and after independence was that India launched a nondiscrimination policy to achieve greater economic development and equality and quality in education. Several prominent Indian educators and philosophers stepped forward to organize India's educational system. Of note are Raja Ram Mohan Roy, Vinobha Bhave, Vivekananda, and others in the late nineteenth century; more recently, in the early twentieth century, were leaders such as Mahatma Gandhi, Jawahar Lal Nehru, Zakir Hussain, Rabindra Nath Tagore, Radhakrishnan; and in the latter part of the twentieth century, mention should be made of Krishnamurti, Shankar Dayal Sharma, and Prem Kirpal, among many others. Mahatma Gandhi denounced the existing system of education for being based on a foreign culture and teaching in a foreign language. He called for a boycott of all government schools and colleges and attempted to institutionalize an alternative system of education. Gandhi's Basic Education policy highlighted vocational learning and the use of Indian languages and served as a foundation for India's Constitution. The scheme of Universalization of Elementary Education (UEE) provided free, compulsory education for all children up to the age of 14 years (Saini, 2000). Some prominent philosophers and educators started even model academic institutions, such as a school by Rabindra Nath Tagore at Shanti Niketan; an educational *ashram* by Aurobindo Ghose in Pondicherry; and a school by Krishnamurti at Rishi Valley. These schools and universities are popular even today.

It is clear from this brief evolutionary account of education that at every stage in India's history there have been prolific writings, initiatives, and reforms by philosophers and educators on educational aims, curriculum, and the role of schools and teachers. The specific names I have mentioned are only a select few of the many people, institutions, and writings that have greatly influenced education in India. Throughout the nation's history, there have been several monarchs and political administrators who have supported, preserved, and promoted learning; furthered education by working on building schools, colleges, and libraries; encouraged public support, respect, and recognition for teachers; and extended a system of indigenous education as much as possible throughout the country. I now provide the reader with a quick overview of the nature of the educational system that is currently in place in India.

The Current System of Education in India

Today, India is among the world leaders in technology, space, nuclear research, electronics, industry, and agriculture, and her educational system is the third most extensive in the world after America and Russia (Saini, 2000). It is India's educated, skilled, and English-speaking masses who have made the nation a prime location for being the backyard for corporate powerhouses from not only the United States but also from many countries in Europe and the Asia-Pacific. Independent India's commitment to the spread of knowledge and freedom of thought is reflected in her Constitution: "The State shall endeavour to provide within a period of ten years from the commencement of this Constitution, for free and compulsory education for all children until they complete the age of fourteen years" (Article 45). Although education is a responsibility of the country's central government, both central and state governments play a major role in the development of education, especially at the primary and secondary school levels. By the 1990s, there were more than 888,000 educational institutions in India with an enrollment of about 180 million. This included about 598,000 primary schools, 176,000 elementary schools, and 98,000 high schools (National Council of Teacher Education website, 2001). The elementary educational system was the second largest in the world with an enrollment of more than 149.4 million children in the age group 6–14 years and employing about 3 million teachers (Government of India's National Policy on Education, 1992). By 1995, there were about 4.3 million teachers working at different levels of school education. Today that figure is much higher.

India's educational system continues to be driven by a strong focus on the assessment of students through a rigid examination system. Government organizations such as the Central Board of Secondary Education (CBSE), which is the world's largest educational board at the operational level, promote a national curriculum that is translated into a prescribed syllabus and corresponding textbooks for each grade level. In addition to the CBSE, some of the other curricula followed by schools in India have been designed by organizations such as the Indian Certified School Examination (ICSE), State Board of Examinations (SBE), and the Matriculation Board of Examination (MBE). Of these, the most widely used in urban schools are CBSE or the ICSE. Students are individually assessed at the end of each academic year for the content of the textbooks they have studied. The policy on language of instruction in all schools follows the three-language formula adopted by the Education Commission in 1964. In each state Hindi is taught by virtue of being the central national language of India; English is taught for being the official and business language; and the third language taught is the native or regional language of the state, 1 of the 22 dominant regional languages recognized as official from the 845 spoken in India. In support of this multilanguage system, the Constitution requires all states to publish books in up to a dozen or more languages (Saini, 2000). Given the extent of diversity in India, the system of education is a complex multitiered system that encompasses an infinite variety of schools. Apart from the CBSE, each state has an independent role in defining the details of education offered in its schools. There is a large system of government schools throughout the country, but also an extensive system of private schools, especially in urbanized areas, catering to India's massive middle-class population of 350 million. Children of poorer families from low socioeconomic classes and rural areas usually attend government schools as these are affordable, being free or subsidized, whereas private urban schools with high tuitions are more accessible to the middle-class population. According to Varma (1999), India has the largest number of out-of-school children in the world, who come mostly from poorer rural households, but at the same time India also has the largest pool of trained and skilled manpower in the world that reflects a largely middle-class urban background. Recent news reports indicate that with increasing awareness due to the current boom in Indian technology and economy, poorer families in India are becoming increasingly dissatisfied with the bureaucracy, overall quality, and poorer administration of government schools and are turning more and more to private schools in the hope that their children will attain quality education and skills in the English language ("India's poor bet precious sums," 2003). The result of this is the mushrooming of private schools and a diminishing number of government schools in small and large cities and urban centers all over the country. Since education is such a priority in Indian society, a large percentage of high school graduates go onto college. At the same time, because of the overwhelmingly large population and a subsequent acute job shortage, there are millions of college graduates who do not have jobs.

I now shift the focus specifically toward early childhood education and present an overview of the education of young children in India.

Scope and Diversity of Current Early Childhood Education in India

It was the establishment of the National Policy on Education (NPE) in 1986 and its subsequent modification in 1992 that accorded a great deal of importance to Early Childhood Care and Education (ECCE) and recognized it as playing a critical part in the development of human resources. Development of ECCE programs was seen as not only providing the individual child with early care and education, but also releasing women for other activities besides childrearing, and facilitating the access to schools by older girls who would have been otherwise providing sibling care. The predominant system of education in most government and many private schools has followed, until recently, the pre-independence, traditional colonial model characterized by a structured approach with a tightly prescribed content-based curriculum; a strong focus on reading, writing, and arithmetic dominated by rote learning; children seated most of the time at individual desks placed in rows; and a large amount of homework assignments required to be completed by the children everyday. At the early childhood level, there is a wide range of schools, and differences may be attributed to whether the school administration is government controlled or privately controlled; whether schools exist in rural areas, small towns, or in large urban centers; whether they provide services as day care centers or nursery schools; whether they are religious or secular in nature; and so forth. However, according to the Sixth All India Educational Survey in 1996, the total coverage of young children in all kinds of pre-primary schools was only 25 percent of the total number of children. In general, the following kinds of pre-primary or early childhood settings may be seen.

The most common kind of early childhood care and education centers are the *Anganwadis* run by the centrally sponsored Integrated Child Development Services (ICDS), and the *Balwadis* run by the state government or local bodies. The goal of ICDS is to empower underprivileged children younger than the age of six years, and ensure that they are physically healthy, mentally alert, emotionally secure, socially competent, and intellectually ready to learn when they reach primary school age. A report by the National Council of Educational Research and Training (NCERT) in 1999 revealed that more than 70 percent of *Anganwadis* had activities for children that were geared toward rote learning and were teaching the 3Rs or reading, writing, and arithmetic. In an exemplar model of *Balwadi*, a member of the community would be selected and trained for the teacher's job, indigenous materials that are easily available would be utilized, and extensive community participation would be encouraged; this would compensate for the lack of funding, resources, and teacher training (Gokhale, 2005).

Another setting is the NGO-sponsored early childhood center. NGOs have played a critical role in advancing childcare services in India, with their unique capacity for mobilizing local communities to bridge the gap between the government and the people. NGO-sponsored education centers may be found primarily in low socioeconomic urban settlements and in rural districts.

The private nursery school operating independently is a common sight in urban residential centers. Usually this school is started as a business by stay at home middle-class wives and mothers, either in a portion of their own homes or in a separate residential building. Such schools are small and most often offer a curriculum to prepare children to be admitted to larger private schools.

The nursery or kindergarten grades within a larger, comprehensive, private school in large urban centers and metropolitan cities comprise yet another type of early education program. Usually, the private schools hold parent interviews and admissions tests for the children, and admission into some of the more elite schools is very difficult.

The mobile crèche, which is essentially a temporary structure set up at a construction site to care for the children of women workers, is run by the

Central Social Welfare Board. The children cared for are usually zero to three years of age, with the primary objectives of the crèche determined by the child's biological rhythm, alternating sleep, feeding, play, and rest.

Yet another type of setting, the urban day care center, is usually run by voluntary and governmental organizations such as the Central Social Welfare Board for disadvantaged families.

In small rural villages, one can see a school under the trees with an adult member of the village taking on the teacher's role and instructing a group of children in the outdoors, equipped with only a portable chalkboard and chalk; the children use slates and chalk as substitutes for notebooks and pencils.

Among these diverse early childhood settings listed, those run by the ICDS (including the *Anganwadis* and *Balwadis*) form the largest group for children in the age range zero–six years. ICDS is considered to be one of the world's largest integrated early childhood education and care program, which serves more than 2.4 million preschool children (Saini, 2000). The goal of ICDS is to not only empower young children, but also empower women in the age group 15–45 years, thus promoting school readiness in children and basic education skills in mothers.

The 2002 National Conference on Early Childhood Care and Education organized by the Department of Elementary Education at NCERT reports that although in 1999 there were 22.9 million children in the zero–six years age group receiving Supplemental Nutrition Program services, the main area of concern of preschool education offered by *anganwadis* and *balwadis* has been quality, specifically with regard to lack of resources; lack of awareness; lack of indicators; distortions in curriculum; poor training; lack of institutional capacity; and inadequate advocacy about the need and significance of early care and education programs (NCERT, 2002). Although the purpose and philosophy of these educational initiatives is good, the resources are scarce and the training of staff is inadequate. The private nursery schools and kindergartens, on the other hand, may be financially supported by high tuitions and/or other private funds, and definitely have the resources to offer better facilities and services, as well as hire teachers who are better qualified and experienced. This can have negative and positive consequences. The competition and reliance on private tuitions may lead to higher academic standards but may result in overcrowding of classrooms and social inequity.

A report by the State Council of Educational Research and Training (SCERT) indicated that although early childhood classrooms have been defined by a structured traditional approach (SCERT, 1994), the National Policy on Education in 1986 recommended a more play-based approach for both primary and pre-primary classes under the influence of the Western discourse of progressive early childhood education. However, a study in

1992 revealed that only 15 percent of early childhood centers practiced this philosophy (SCERT, 1994). In light of the above, recent years have witnessed initiatives for change in early childhood and elementary education within both the private and government sectors, attempting to make the approach less rigid, more child centered, more activity based, more experiential and creative, and less structured and formal (Gupta, 2001; Kaul, 1998; Thapa, 1994).

The Hindu Philosophy: Pervasive and Practical

It is interesting to note that in spite of the changing models of mainstream education under the influences of various historical and cultural factors, the basic values and beliefs that children in India have been taught formally and informally remained somewhat constant, drawing from an ancient Hindu philosophy that prescribes a way of life and continues to be a part of the country's philosophical and spiritual discourse. Even today, the same values and beliefs are upheld and practiced by many Indians in their daily lives, and children are taught these values in many different contexts: informally at home by their parents, grandparents, and other members of the extended family; through the "hidden" curriculum in their schools; via social agencies such as cultural centers, institutions for prayer and worship, organizations promoting environmental preservation and conservation issues; via media channels such as the television, radio, cinema, and print; through popular and classical music, songs, and dance; through theatrical pageants and dramas based on historical events, classical literature, and mythology that are enacted during the celebrations of all holidays; and through written and oral forms of folk tales and traditional stories for children. These values and beliefs remain very much alive in the language, customs, rituals, celebrations, and ceremonies that mark births, deaths, marriages, and other significant milestones in life. Today, subsequent to a colonial encounter, the urban Indian way of life reflects age-old values based on a worldview that is deeply rooted in the teachings of the *Veda* and *Upanishad* and in the concepts of *dharma* and *karma*; along with the modern thinking patterns and lifestyles of younger societies and cultures in Europe and the United States. Despite great diversity among the people in India, there is a certain Indian identity, or Indian-ness, which results from the fact that key aspects of the *Veda* teaching a way of life have worked their way into the ethos of the larger culture and society in India. Max Mueller, the well-known scholar and Sanskritist, noted in 1882 that there was an unbroken continuity between the most modern and the most ancient phases of Hindu thought (cited in Nehru, 1991, p. 88), and that observation is valid for Indian society even today.

Over the past 30 years or so, schools in India have taken on a more Indian flavor and have rapidly lost their "British" feel. Stories from the Indian epics, mythology, and history, as well as Indian classical and folk music and dance have surged into textbooks and the curriculum of most schools, and a more deliberate emphasis is being laid on the teaching of Indian values. There is a movement toward the philosophy that is based on ancient texts such as *Veda*, *Upanishad*, and *Bhagvad Gita*; this movement is restricted not only to nationalists, but appears to be supported by educators, philosophers, and a large percentage of the general public, as is manifest in articles, dialogues, and conversations that are covered by the media ("Kindling the Vedic spirit," 2000; "Need to tap Vedic knowledge," 2000; "Education is key to spirituality," 2002; "Desperately seeking another Vivekananda," 2002). There is a widespread desire to reclaim intrinsic Vedic thoughts and offer an educational experience to children that will be culturally relevant, values based, as well as progressive in nature.

One of the focus areas identified in the 2001 National Curriculum Framework for School Education prepared by the NCERT is "broad-based general education to all learners . . . to help them acquire basic skills and high standards of Intelligence Quotient, Emotional Quotient and Spiritual Quotient" (NCERT, 2001, p. 34). The framework emphasizes a need for strengthening national identity and sends out a strong plea "for promoting national integration, and social cohesion by cultivating values as enshrined in the Constitution of India through school curriculum . . . the ten core components being the history of India's freedom movement; the Constitutional obligations; the content essential to nurture national identity; India's common cultural heritage; egalitarianism, democracy and secularism; equality of sexes; protection of the environment; removal of social barriers; observance of the small family norm; and inculcation of scientific temper" (NCERT, 2001, p. 36). It is important to note here that the word "secular" as used in the Indian context implies an equal acceptance of all religions. This is contrary to the meaning of "secular" in the United States, where the term implies a rigid separation between church and state leading to an absence of religious observances in schools. The document published by NCERT also specifies that "while, on the one hand, education should help in promoting a global world order, on the other, it should be seen as developing a national consciousness, a national spirit and national unity essential for national identity" (NCERT, 2001, p. 12).

As previously mentioned in this chapter, earlier attempts to ground education in Indian philosophical teachings before and after the independence of India in 1947 led to the establishment of renowned institutions founded by philosophers such as Rabindra Nath Tagore at Shanti Niketan, Krishnamurti at Rishi Valley, and Aurobindo Ghoshe at Pondicherry, among others. However, although these schools were (and continue to be) regarded as excellent, they draw on a specific pedagogy and spiritual philosophy. Consequently, they are situated away from mainstream education, and have had to train their own teachers to implement their educational goals. An important point to note is that although modern leaders and educators of twentieth-century India, such as Lajpat Rai, Tagore, Gandhi, Nehru, Radhakrishnan, and Sri Aurobindo, agreed that the goal of Ancient Indian education should remain the goal of contemporary education in India, they also emphasized the importance of simultaneous scientific and technical training for modernizing the country (Gotz, 1995). More recently, Dr. Reena Ramachandran, director general of Fortune Institute of International Business affirmed this balance, emphasizing that Indian society can experience an erosion of its value system if there is an inadequacy in the growth opportunity in India's economy (Woman of the Year, 2002). There appears to be an intrinsic understanding within the Indian worldview that maintaining traditional values and embracing modernization do not have to be mutually exclusive.

In the next chapter, I illustrate how some of this is reflected in an actual school setting in urban India and in the classroom practice of early childhood teachers, specifically those who participated in my study. Before I bring their voices out into the playing field, I would like to give the reader some idea of who the participants were. The research was a naturalistic, descriptive inquiry, and my fieldwork involved different modes of data collection including interviews, surveys, classroom observations, and document review. I selected a variety of participants who included schoolteachers, school principals, and teacher educators. The findings and conclusions for the study were based on the responses received from 35 early childhood teachers, 13 school principals or administrators, and 2 teacher educators. The majority of the participants took part in a survey, and although their responses are reflected in the overall findings, the voices in this book represent only those specific participants whom I interviewed. Charu, Vasudha, and Malvika were the early childhood teachers who shared their experiences and classrooms with me and whose voices are heard most often in this book; Shakti was the principal of the school where these teachers worked; Antara was an elementary school headmistress; and Veera was the supervisor of six nursery classes. In addition, I interviewed two teacher educators, Nagma and Ashwini, from the two teacher education colleges that Charu, Vasudha, and Malvika had attended. So it is primarily the voices of these eight participants that the reader hears during the course of the book.

Chapter 4

Aims of Education Contextualized within Urban Indian Society

The school practitioners whom I interviewed and whose voices I share worked at The School, a large private school that is typical of the "public schools" in India. Public schools in India refer to independent schools with high tuition costs and fees as opposed to U.S. public schools, which refer to state-run tuition-free schools. The School was housed in a large red brick building and at the time of my visit a fleet of yellow school buses stood parked outside the gates since it was close to dismissal time. The sun shone brightly on the hundreds of potted Chrysanthemum plants that were lined up against the brick walls. There were the yellow, pink, mauve, and white blossoms that are typical of Delhi in the wintertime when one can see them everywhere—full, fragrant, and expensive. The grounds of the school were neatly manicured and very picturesque. Near to where I stood was a rockery decorated with small, white, marble cherubs and a tiny cascading waterfall. In another section of the grounds was a Japanese garden with a small arched bridge and more figurines. The plants were green and well maintained. All the classrooms in this school were similar in design. The second grade classroom I visited that morning was furnished with only tables and chairs. There were 20 tables that seated 2 students each, and a total of 40 chairs. A blackboard (chalkboard) was mounted on one wall of the room, and in front of it was placed an adult sized desk and chair for the teacher. Another wall had windows that overlooked the front grounds of the school. A third wall was equipped with a free standing closet in which were usually stored children's notebooks and classroom supplies such as chalk, pencils, pens, and the like. The other three walls of the hexagonal classroom were lined with bulletin boards that displayed colorful charts and posters. Several three dimensional projects had been placed on the floor and on the windowsills. I was told that students had been working on these projects for an assignment that was related to the environmental studies program in their school curriculum.

I focus now on the aims of early childhood education professed most frequently by the teachers, school principals, and teacher educators who participated in my study, and discuss why these educational aims are considered to be important by positioning them within the larger context of education in India. During the course of the study, questions had been asked of the participants to describe what, in their perception, was an ideal education and what was most important for young children to learn. Their descriptions indicated that the aims of education prioritized by them were:

- teaching of values and right attitudes;
- intellectual development and developing the ability to think;
- developing proficiency in academic skills; and
- developing the recognition and support for cultural and religious diversity.

I take up the discussion of each of the four educational objectives individually, and explore how they might be influenced by sociocultural and historical sources, and trace their relationships to concepts that are basic to a way of life commonly practiced in Indian society.

Aim 1: Teaching of Values and Attitudes

Judging by the responses from teachers who completed the survey as well as the three teachers I had interviewed at length, it seemed there was wide consensus that teaching values and good attitudes should be one of the most important aims of education for young children. Each teacher reiterated this point several times. Charu taught Class Two (second grade) and had 40 students in her classroom. Prior to this she had also taught Kindergarten for a few years. The reason that she chose teaching as a career was because it would enable her to spend more time with her own daughters and she wouldn't be forced to leave them alone at home or in the care of a servant even for a short while, and this was very important to her. She proceeded to articulate what she felt was the most important goal of education:

> See, basic values are very important, and a very good awareness of their surroundings, their responsibilities to the environment, to their society.

These I think are the most important things . . . Responsibility to society is the most important thing . . . basically, to be able to adjust themselves to the surroundings, and as they grow up how they should relate to their friends, to others in society . . . These things, I feel, are the most important things that I would try and inculcate in my students.

Charu saw this aim of teaching children the right values such as responsibility, consideration for others, and appropriate social behaviors as being of utmost importance. She emphasized that careers and professions were increasingly being chosen these days for the ability to make money, and she personally made it a point to keep reminding her own children as well as her students, "Don't just think of making money. I want you to do something for people also. Helping people, making somebody's life better." She felt it was important to teach the children these values because she explained, "I *want* them to be better human beings and stronger people." Encouraging children to adopt the right attitude in difficult situations was an important educational responsibility for Charu:

The way you project yourself, or the way you react to situations, is something I'm very very [double emphasis] particular about. Even my kids, I keep telling them at home and in school also . . . There might be times when things are not according to the way you want it to be. One must be cool and patient at such times. And it is in such trying times that it's a true test of your character . . . If someone hits you . . . try and control yourself, reason with the person . . . I want you to behave in the right manner and do what is right.

She summed up her thoughts by saying, "so these are the more important values—being honest and upright and straight. And being a good human being."

The second teacher I interviewed, Vasudha, had been a Nursery teacher for several years and had a daughter who was similar in age to her students. She knew she always wanted to be a teacher because she viewed it as a career she could pursue without having to sacrifice the quality of her family life. She had 40 four-year-olds in her classroom and like Charu emphasized that the primary goal of education should focus on the teaching of values:

Look, education is just not learning A-B-C-D and 1-2-3-4. That, I think, comes later on. But initially what the foundation has to be laid on is good values, and for the child to adjust very well to the school routine, to mix around with other children and interact in the best way . . . And then the best values should be inculcated. That I think is of the utmost importance . . . all the values should be inculcated, the child should have an all-round personality development, not just academic.

Malvika was the youngest of the teacher participants in the study and was a fresh graduate from a teacher education college. As a novice teacher, this was her first year of teaching and she had been assigned to Class One (first grade). Like the other teachers in the school, she also had 40 students in her classroom. Independent of the others, she too affirmed what her colleagues had expressed regarding the primary goal of education for young children:

> See, the most important aims of education should be to give them values, moral values. To develop certain courtesies in them so that whenever they grow up they are basically good human beings. Academics I only give fifty percent weightage and the rest is all non-academics . . . their personalities should be such that wherever they go, whatever they do in life they should be successful. This is the main aim of education according to me . . . moral education, value education . . . things which will make them more confident. Academics [alone] will never make them confident.

According to Malvika, it was the teaching of morals and values more than the teaching of academics that developed confidence in a human being. She shared with me that she herself had been very good in academics as a schoolgirl, but being a gold medalist alone had not helped to make her a confident human being. Confidence was developed in the children by helping them become aware of their responsibilities and motivating them to do their best. She described how she approached this objective in her own classroom:

> [W]e give them small projects, small responsibilities, like we've got monitors for the week—this week, whenever we're going out of the class this is the person who has to remember that all the lights and fans have to be switched off . . . First, they learn why we switch off the lights and fans when we're going out. Secondly, the child learns to handle the responsibility—every time we go out he's the one who will take the initiative and will switch off the lights and fans. Or we've got Distribution-in-charge for the week, or Cleanliness-in-charge for the week. And they want to be at their best because . . . if they do their best, next time again they get a turn . . . It gives them a good self-concept.

Malvika, in addition, articulated the importance of acknowledging the personal values of students in the classroom and being sensitive to people's ways of life. Her reasoning was grounded in her own school experience. She shared a touching incident that occurred when she was a student in Class Eleven (eleventh grade). She described how one day one of her male classmates came to school with henna on his hands. In India, the paste from dried and ground henna leaves that have been soaked overnight is commonly used by girls and women for decorating their hands on various occasions. The teacher commented on his hands and the fact that only girls used henna. The boy was terribly embarrassed and tried to explain that his sister had just got married and in their family it was customary and traditional that the sister could not marry until her brother had had his hands painted with henna first. The teacher was insensitive to the context of this family tradition and this led to snickering in the classroom. Malvika remembered agonizing over this incident because although she wanted to she could not sympathize with her classmate openly and had to force herself to snicker along with the teacher and the rest of the students.

> [W]hy are we shy about our own values . . . see, as a student this is how we learn to accept each other. And we have to develop that acceptance in the teachers and other students. And I think this stage has to be initiated right from early childhood . . . You know, if you are not going to respect your own values how do you expect others to respect your values . . . If I'm scared or shy to show my values then how will you accept that? . . . I think teachers play a very important role in giving that sort of a confidence to the children.

During the course of the interviews, I asked certain questions that provided opportunities to the teacher participants for explaining and defining what they meant by the values to which they were referring. Many of the so-called values turned out to be universally recognized virtues such as honesty, kindness, sharing, loving, caring, and so forth. Other values that had been described were such things as getting along with people, being able to relate to friends, and being able to adjust to their environment. Still others included helping students develop a sense of responsibility, initiative, independence, and a good self-concept. Many of these are concepts that might very well be included by definition in the domain of child development commonly known in the West as socio-emotional development, It is interesting to note that these concepts were being termed "values" by teachers in India. This itself is an indicator of the "value" or importance attached to these learned behaviors in India in terms of how the individual is viewed in relation to a larger group outside of himself, whereas the same learned behaviors in the West are grouped under socio-emotional development, which indicates an inward psychological look into the development of the individual in relation to himself/herself.

Antara, an elementary school headmistress, also highlighted the importance of teaching values and right attitudes along with other skills and knowledge:

> The curricular as well as the co-curricular activities of the school attempt to transmit the norms, beliefs, values, attitudes, behavior, knowledge,

technological, social skills, etcetera of the society to the new generation . . . controls and regulates their behavior with regard to the core areas of social life but also prepares them to be able to adjust to change.

During my conversation with Ashwini, one of the teacher educators, I brought up the wide-spread emphasis with which the teachers were reinforcing values in their classrooms and asked her what she thought of that. She responded by saying, "I'm willing to understand that. Because that whole—the *guru shishya parampara* [the teacher–student tradition of Ancient India] and the cultural backdrop in that is so strong that I'm not surprised." Her response implied that she found it perfectly natural that teachers would focus strongly on values defined by the Indian culture in their classroom practice.

The responses of many other teachers to the more extensively distributed survey also indicated similar results. Of the 42 early childhood teachers and school principals from 15 different private schools in New Delhi who responded to the question, 90 percent indicated that teaching of values was either the most important or the second most important aim of education. In order to illustrate more lucidly what the participants meant by "values," I sorted through the values that had been specifically mentioned by them during the interviews and on the survey questionnaires and found they could be placed under three categories: common virtues that are found more or less universally across cultures; values that are specifically mentioned in Indian texts and scriptures; and skills that are named and recognized as appropriate by child development theories. Figure 4.1 illustrates the three categories with the specific values that were mentioned by participants as important educational aims.

In attempting to capture the perspectives of some Indian early childhood educators in New Delhi by contextualizing their practice, and in trying to understand and explain the forces in the field that defined those practices, I have tried to be as objective as possible in my interpretations and discussions of the findings. Although it had not been my intention to develop an extensive comparison between early childhood practice in India and in the United States, some degree of comparison became relevant because the findings did highlight the discrepancy between a teacher education curriculum created under Western influence on the one hand, and the actual early childhood classroom curriculum informed to a large extent by the Indian values and belief systems of the teachers, on the other hand.

These "values" appear to be consistent with many of the concepts included in the discussion on Indian philosophy in chapter 2. I would like to remind the reader to view Hinduism more as a way of life than a religion. The *Veda* and other scriptures such as *Upanishad* and *Bhagvad Gita* offer detailed recommendations for a good way of life and living, indicating that

Universal Virtues	Indian Values	Developmental Skills
learning to love humanity	llearning to respect community living	learning to work in teams
learning how to share	developing a strong character	learning good social skills
developing confidence	learning how to get along with peers	developing independence
developing strength		developing individualism
learning to have faith	learning to live together	developing social awareness
being truthful and honest	learning to face the world without fear	developing positive attitude
learning to have trust	learning to respect elders and others	learning analytical skills
learning to be kind	learning to be dedicated toward work	learning logical thinking
learning to be caring	learning to respect diversity	developing self-esteem
learning good citizenship	learning to understand the idea of Karma	developing good self-concept
learning to be a good friend	developing a healthy personality	
learning to be polite	exhibiting overall good behavior in life	
developing integrity	learning to respect one's culture	
developing compassion	learning to be tolerant of different religions	
learning to be responsible	learning appropriate attitudes and norms	
	learning humility	
	learning good moral values	
	learning to persevere, never giving up	
	learning to accept failure as a friend	
	learning appropriate classroom behavior	
	inner strength to face difficulties in life	
	developing a secular outlook	
	being conscious first of one's duties and then of one's rights	
	developing holistically and harmoniously	
	learning to care for the elderly	

Figure 4.1 Values articulated by teachers as being important aims of education

a life lived within one's *dharma* and mindful of one's *karma* is the way to achieve salvation. Thus it is possible to be a good Hindu without being religious or ever going to a temple. The distinct break-up of Hinduism into religion and philosophy as two different compartments became prominent during colonialism under the Greek-influenced approach to philosophy in the Western world. Earlier, in my discussion on the conceptual understanding of the self and of moral development within the Indian worldview, the all-pervasive idea of *dharma* was explained as being central to Indian cognition and the Hindu worldview. The meaning of *dharma* cannot be easily and accurately translated in English, but it implies several ideas in juxtaposition: morality, ethics, responsibility, duty to others. It is what binds a society together and maintains a harmonious existence between members of a community. The concept of *dharma* is not merely a philosophical abstraction, but its actual practice in day-to-day life is what every Indian strives for. Indians place a high degree of importance on ideas related to *dharma* such as respect for elders and carrying out one's duties and responsibilities to society. Further, these duties and responsibilities are defined by the multiple roles that each individual plays within a society. It is understandable then that the teaching and reinforcing of values that are connected to moral development and *dharma* become an important aim of education in Indian classrooms, and responsibilities of the individual become prioritized over the rights of the individual. Defined more by a spiritual and cultural philosophy and not so much by their pedagogical training, teachers' own upbringing, background, and system of prior beliefs and practical knowledge influences them to identify values as an important educational aim. According to the ancient Vedic philosophy, education's final goal was to uplift human character through a process of self-renewal and self-development (Nanavaty, 1973; Vyas, R.N., 1981), and the educational objective throughout India's history has been that noble ideas be enshrined in the human heart. Consequently, Indian educational psychology has been value oriented, and it seems to continue to be so as is evidenced from the responses of the participating educators.

At a conference in January 2002 of the NCERT, India's premier government organization for developing a national curriculum, syllabus, text books, and teacher education, the Dalai Lama delivered the keynote address and noted that the absence of human values such as love, compassion, forgiveness, sharing, and caring makes a person unhappy and nurtures feelings of insecurity and suspicion. According to a newspaper report, the Dalai Lama was full of praise for India's ancient *guru–shishya* (teacher–student) tradition, and recognized that whereas modern education was more secular in its outlook, traditional education combined knowledge of the world with knowledge of the soul ("War must never be," 2002). The acknowledgment of Ancient Indian beliefs by the Indian government is also reflected in the home page of the Government of India's website for the Department of

Education, which displays a full screen graphic of an *ashram*, or a forest school of Ancient India, with the students, their hands folded in prayer, seated respectfully on the ground around their teacher, or *guru*, who is himself in the midst of deep meditation. The national flag of India displays the *chakra* in its center, or the wheel of life, which symbolizes the cyclical notion of life and death central to Hindu belief. These are just a few examples that illustrate the pervasiveness of ancient Vedic philosophy in Indian society, its strong connection to general education, and its clear visibility in day-to-day life.

Aim 2: Intellectual Development and Developing the Ability to Think

An educational aim rated very highly by the participants was intellectual development and developing the ability to think. The ability to think and the development of the intellect assumes priority in Vedic philosophy also and has historically shaped the concept of education in India through scriptures, mythology, science, and literature. Shakti, the principal of The School stated unequivocally:

> . . . most importantly I would give the highest priority to creative thinking skills as the biggest role of formal education in a school . . . good education for me should lead to making a child think. By thinking I mean, think right. Because life is an understanding of paradoxes and opposites—making the right choice. Everything has an opposite—which road to take, which clothes to wear . . . Right choices is only one part of thinking . . . It requires logic, it requires understanding, it requires relationships, it requires language . . . his [the child's] mind should grow and not be stumped by too much authority . . . The inner soul must open up. I would call it an awakening.

After my formal interview with her was over, Shakti in fact returned to this topic and shared further reflections on it. She felt that without the ability to think clearly, nothing was really possible because it was this ability that really drove and gave rein to the emotional, social, and spiritual skills of an individual. And it was the ability to think that helped an individual cope with the rough patches in life. Shakti emphasized that emotional skills could not come into play without thinking skills and spoke about the importance that had been given to the clarity of intellect and thought in Ancient Indian texts saying: "Absolutely, that was the most important thing" and that is what education needed to do—help develop the intellect.

Veera, as a supervisor of nursery teachers, also articulated the educational aim for four-year-olds as the ability to think but couched it in different imagery. She defined teaching to have been a success if "we have taught them not to be afraid of learning . . . we do not dampen their curiosity—and if by the end of the year I have children who are curious, who are not afraid to ask questions, and who look at coming to school as a fun thing." The two teacher educators who participated in the study also shared their opinion that the important aim of education was to develop the ability to think. Ashwini used the words "awareness and surprise," explaining that these terms translated into "training people to think, to question, and to act." Her colleague, Nagma, a veteran teacher educator until her retirement a few years ago, conceptualized education as a process with the aim of developing:

> an autonomous person . . . somebody who can think for herself, and knows how to find resources which will help her to sort out problems, somebody who can use other information but not necessarily be absolutely dependent on other people—somebody who is reflective . . . And allow children—and teachers—to organize knowledge for themselves—to facilitate it . . . giving people a voice, letting people think for themselves and breaking down—you know . . . systems . . . of hierarchy and . . . divisions.

The emphasis on this educational objective also has roots in ancient Vedic philosophy. In chapter 2, I had drawn attention to the importance that was given to knowledge and the intellect in education in Ancient India. The very word *Veda*, the ancient scriptures upon which all of Indian thought and philosophy is based, originates from the Sanskrit root "*vid*," which means "to know." In addition, this philosophy views Reality not as something that can be seen or touched, but that which can only be experienced and understood by the mind, thus placing a high value on cognition, intellectual knowing, and comprehension. For the purpose of providing a point of reference, I have taken the definition of the term "intellect" as meaning to be "the power of thought and mind, the ability to reason or understand or perceive relationships and differences" (*Webster's New World Dictionary*, second college edition).

Educators in Ancient India laid great emphasis on the acquisition of knowledge to enlighten the mind. Knowledge did not refer to just a body of information but included the recognition of relationships, cognition, comprehension, and so forth. Ancient texts suggested that the learning process depended on factors such as having a desire to learn; receiving of information; understanding of the facts and then retaining the facts in one's memory; reflecting upon what one had understood; exercising judgment on what one knew; and developing the power to discriminate between various bodies of knowledge (Achyuthan, 1974; Vyas, R.N., 1981). Interestingly, these processes follow an order similar to the modern day understanding of the taxonomy of cognitive thinking developed by Bloom in 1956 namely: knowledge, comprehension, application, analysis, synthesis, and evaluation. So the primary goal of education in Ancient India became that of helping students to hone and refine their skills of the intellect; to help them retain knowledge and information; to help them understand, comprehend, and reflect on that knowledge; and to help them learn how to differentiate based upon what they knew. Achyuthan (1974) also specifies that the importance given to learning and knowledge influenced the factors that were said to bestow honor and respect on families and individuals, with erudition being at the top of the list, followed by work, age, and material wealth. This form of classicism is manifest in parts of India even today in the idea that an individual who belongs to a professionally educated family is accorded more respect than one who belongs to a business-class family. I believe that the reason for this might lie in the understanding that education for a profession such as teaching, medicine, engineering, and so forth has generally developed a set of values that are defined by service to others rather than self-gain, whereas an individual from a family of traders and merchants may have acquired a set of values that are more materialistic and are defined by monetary wealth and personal profit. I would like to add, however, that there has been a shift in this perspective in urban metropolitan India in light of recent economic policies that have swept the nation into the realm of global trade, commerce, and business, and capitalist values are becoming more visible and prominent. But clearly, intellectual development was seen to be one of the more important aims of education, and judging by what the participants professed it continues to be so. With honor and intelligence so closely linked to the development of the intellect it is not surprising that one of the most important and frequently recited ancient *Vedic Mantras* in India is the *Gayatri Mantra*. This short verse means: "I adore the Divine Self who illuminates the three worlds—physical, astral and causal; I offer my prayers to that God who shines like the sun. May He enlighten our intellect." As referred to earlier by Shakti it is this clarity of the intellect that gives reign to all the other skills in an individual, and that without being able to discriminate and think clearly between right and wrong, the right choices for appropriate actions, deeds, and thoughts cannot be made by an individual. That the intellect gives reign to the individual's social and emotional development is a perspective that is certainly supported by modern social-cognition theorists.

Another domain that is actively involved in the teaching of values is moral development. Shakti's views along with Charu's emphasis on helping children conduct themselves with emotional balance and developing a social awareness of their duties to others seemed to be consistent with the Indian concept of moral development. Morality is the primary focus in two of the greatest Indian writings, the *Bhagvad Gita*, which describes Arjuna's moral dilemma, and the *Ramayana*, which is the story of Lord Ram's voluntary exile to honor a vow that his father was forced to make. The epics relate how both men adhere to their *dharma* or duty. Activities related to both the *Ramayana* and the *Gayatri Mantra* featured prominently in the Nursery curriculum in Vasudha's classroom as she described her practice: "We have a prayer . . . *Gayatri Mantra* for the children. Which is very important. And not only the *Gayatri Mantra*, there are some other Vedic mantras also which the children recite . . . I had *Ramayan* Assembly this time and I had written the whole script on my own . . . the show was really good and everyone enjoyed it."

In light of the great importance attached to the first two aims of education it was not surprising to hear from many school principals who participated in the study that teachers were selected with great care; and although institutional degrees in education were a formal requirement, informally more significance was given to the teacher candidates' individual background, upbringing, and personality. Although the teaching of values and development of the intellect as educational aims were not as overtly distinct in the school's prescribed curriculum as was academic proficiency, they nevertheless surpassed the latter in importance to such an extent that many school principals desired to hire teachers who would draw from their own personal belief systems and positive dispositions, and reinforce in their students the Indian values that they themselves were raised to follow. In this manner, the "hidden" and covert nonacademic curriculum was seen to become more powerful than the prescribed academic curriculum in the cultures of many of the private schools that were included in the study.

The next educational aim professed by the participants was ensuring that students achieved a high level of academic proficiency.

Aim 3: Developing Proficiency in Academic Skills

Although the teaching of academics was not high on the list of educational aims that had been articulated by the participants, it assumes a natural priority in India because this goal is inherent in the very system itself. The system of education in India as a whole within which schools, classrooms, and teachers must necessarily function is propelled by exam-oriented, textbook driven academic proficiency, and the teacher participants recognized this as being an important overall goal for the school. However, they also believed that because this was such an important systemic goal it would be taken care of by all who were involved—the teachers, the parents, and the students themselves. The non-negotiability of high academic achievement in India makes it necessary that students be given all the required support in their classrooms, in their homes, and even in out-of-school private tutorial classes to ensure that their academic performance is maintained at a high level. Keeping this in view, the teachers felt that academic teaching would happen in any case, and therefore they needed to make the extra effort to ensure that values are also being taught to the students. Charu articulated exactly this thinking when she said, "academics they will learn. Their parents will make sure because they know that the system is so marks [grades] oriented." Independently, Malvika echoed the same sentiments as she explained that in terms of an ideal education:

> . . . no doubt, we cannot leave apart the academics . . . but academics they'll anyway pick up—there's pressure from home and tutors and everybody, so academics they'll anyway pick up . . . See, writing and all they'll learn. We've got a formal curriculum which we have to follow, and we do it. I mean, if we have to cover a certain amount of course [coursework] this month, we will do it without anybody saying anything to us.

Malvika revealed her technique of integrating what I have termed "meta-academics" into the teaching of academic work in the classroom. She illustrated this point by the following example in which a lesson about the moon, as a scientific and academic topic, is enriched meta-academically through the skillful interweaving of contextual and culture-specific information:

> . . . we can talk about the academic part of it—that the moon is a satellite of the earth and the moon does not have its own light, but we can also talk about how it reflects our star signs, and how we are born on certain days, and the lunar calendar, and how the moon has a special relationship with the tides, and then coming back to the Indian culture of how the moon is considered to be *Shanti ka Devta* [the god of peace and serenity]—it's very cool and whenever you are there in the moonlight you feel very relaxed because it's got that inner strength in it which it passes on to you.

In the teaching of academics, there was an attempt to emphasize and highlight such beliefs and value-laden ideas that underlie the basic Indian spiritual philosophy. This was also a reflection of the important place that astrology has in the lives of most Indians. Almost no major decision is made without consulting astrological charts for auspicious dates and times. Even names for people and businesses are selected after careful study of which letters/sounds of the Sanskrit alphabet will be more auspicious for that specific venture. Another example of teaching beyond academics came from Charu, who explained that each year when she taught her class the unit on Water she made sure that she wove in themes on conservation and preservation so that the children could "learn even little things—not to waste water, electricity, paper. . . . And then I tell them that you must teach your parents! Suppose you see anybody wasting water or electricity, what should you do? . . . So these are the more important parts of my teaching I feel, than teaching just English."

Since Vasudha had a young daughter of the same age as her nursery class students, I asked her to share with me what her expectations for her daughter were in terms of academic achievements, and this is what she said: "I want her to be a Scholar Badge [Honors] student. But I will never force her to be one . . . And I want that by the time she reaches Class Three [third grade] she should be on her own and she should be studying [without parental supervision] . . . I want her to be self-dependent. I really want her to be self-reliant." With regard to the curricular objectives for the academic curriculum in her classroom, Vasudha said quite firmly that she did not force her students to learn how to read and write:

> We just give them a crayon, and if they can write it's fine otherwise we don't pressure them. . . . They just recognize numbers from 1–10, and our main motto is to build their foundation. If their foundation will be strong then they'll be able to learn the other numbers also. So from 1 to 10 [based on the numbers one to ten] maximum activities are done . . . then we move on to [concepts such as] After, Before, One less than, One more than and we do lots of activities on that. Basically we are covering numbers 1–10. And in alphabets A-Z, the small [lower case] and the capital [upper case], and the sounds. And they are encouraged to draw a lot—a lot of stress is laid on drawing because it helps in their motor coordination and motor skills.

When I asked Veera, Vasudha's immediate supervisor, to say something about the academic goals for the nursery children, she acknowledged a broad outline of academic objectives that she hoped the teachers would teach the children by the end of the year, which included a certain degree of gross motor and fine motor function; however, she also stated that children in the nursery class "should know this much of alphabets, and that much of number work . . . values and symbols and sounds of alphabets, and maybe some amount of vocabulary . . . but towards the end of the year we might start asking them to write."

I have mentioned earlier that the teaching of academics is one aim of education that is prioritized in all schools in India due to the exam-oriented nature of the educational system as a whole. In chapter 3, I described how the colonization of education in India and the erosion of Indian thought and language in schools started with the passage of Macaulay's Minute in 1835 when English was declared as the language of instruction in all schools founded by the British government in India. India's current educational system is driven by the same strong focus on the assessment of students through a rigid examination system that had been initiated by the colonial administrators in the mid-nineteenth century (Kumar, 1992). Graduation from high school is solely dependent on high scores in the academically rigorous Board Examinations that all students must take at the end of their senior year. Government organizations such as the CBSE thus continue to promote a national curriculum that is translated into a closely prescribed syllabus, and corresponding textbooks for each grade level are designed to match the syllabus. Students are individually assessed at the end of each academic year for the content of the textbooks they have studied. As each year's academic content is built on the previous year's syllabus, the pressure to study the content of textbooks filters down into the early childhood classrooms also. Workbooks for language and math are published and prescribed for the Nursery classes also (Gupta & Raman, 1994). However, recent influences of the media, Internet, and educational exchange programs on private institutions such as The School have led to exposure to the child-centered approach of progressive education in the United States and other nations of the "developed" world. Subsequently, these school administrators attempted to exercise some control over not letting the early childhood curriculum become overly dominated by an academic syllabus. This was apparent by the manner in which Charu, Malvika, and Vasudha strove to balance an activity-oriented approach in their teaching of academic content and concepts. However, maintaining this control is a huge challenge in the face of high academic expectations for children in all age groups, and further, activity-based approaches may not be the norm in other private schools or in most government schools that have been less exposed to the Western discourse of child-centered early childhood education.

Although rigid academic learning characterized by memorization and assessed through strict examinations was not always a characteristic feature in the Indian educational system, the concept of memorization has been historically associated with learning and knowledge acquisition because of the nation's strong oral tradition and the oral rendition of scriptures, literature, mythology, science, history, and folklore. According to ancient texts, memory itself was considered to be an important and innate attribute of the human mind, and viewed as essential for the development of identity of the self as related to the past, present, and future. It is quoted in the *Chandogya Upanishad* that it is "through memory we know our sons, through memory our cattle." According to other ancient texts such as the *Nyaya Sutra*, the development of memory is a four-stage process (Vyas, R.N., 1981). These four stages are: (1) the mind is fixed on the object to be remembered, while the individual tries to focus on the specific characteristics of the object; (2) the individual then mentally tries to organize the characteristics of the object; (3) the object to be remembered is then associated with some known fact or event; and (4) the individual should try to recall the object repeatedly. Memory played an important role in both the understanding and retention of the content. The importance given to memorization in Ancient India is an interesting parallel to current recommendations of medical research for active participation in mental activities to sharpen memorization skills, which could help delay the onset of neurological disorders such as Alzheimer's.

It is important to note here that there appeared to be a definite distinction between academic memorization and intellectual development in the minds of the participants, and both were seen to occur in the curriculum of the classrooms that I observed. As the above examples from Ancient Indian texts indicate, memorization has been considered as a necessary step toward cognitive knowing, the learning and comprehending of concepts, and the retaining of important factual information. This acceptance of memorization as a necessary step in the learning process is an approach that is vastly different from that in many Western educational contexts, where memorization has been strictly discouraged in classrooms. Rote memorization of facts and information assumed great importance in schooling in modern India after the colonial implementation of the stringent requirement for students to pass a series of examinations in order to graduate from high school. The systemic focus on examinations in the educational system in India continues to result in such memorization; however, Viruru explains that memorization does involve a definite engagement of the mind because all the time an individual repeats information, the subconscious is pondering over and processing what is being repeated, and further, one cannot think about what one does not know. This kind of learning style may be the only way to address the educational practice in India, which is so test driven and exam oriented.

One significant observation that I made with reference to the aims of education in the early childhood classrooms I observed was the low level of attention given to sensory development. Although this is a critically important educational aim for young children in the Euro-American context, it did not appear to be a prominent goal in many of the early childhood classrooms in New Delhi. In general, schooling in India has not attached much importance to sensory education, and there may be two reasons for this. First, children are many and resources are scarce; second, in terms of space, time, and money, it would be difficult for many schools to provide basic openended classroom materials that foster sensory development such as sand, water, finger paint, collage material to explore textures and colors, listening stations for auditory development, and so forth. Although this does not seem to be a strong enough reason because many of these experiences can be substituted to some extent with materials found in the natural environment, the large numbers of children, scarcity of running water, unreliable supply of electricity, and lack of adequate indoor space in inclement weather are definitely major constraints. The second, perhaps more significant, reason might be the fact that in Indian philosophy the concept of *maya*, that which can be seen and touched is an illusion, is very powerful. Reality is understood as being something that cannot be seen or touched but rather understood by the intellect. This might explain why very little importance has been given to sensory development in the classrooms and more emphasis has been laid on intellectual development.

The strong focus on the learning of academic skills in the Indian classroom might be viewed rather critically by many progressive early childhood educators in the United States. Nagma, one of the teacher educators, strongly criticizes the focus on academics as being too severe: "But you see the whole system of our education is now promoting high competitiveness . . . Here there is very little space for it [cooperative work] in schools. Children working together . . . that's a very minuscule part . . . it's how many *marks* you've got. And that's what's being promoted." Ashwini, another teacher educator, presented a different perspective. She drew upon her own background in classical Indian dance and made a point of carefully describing how classical dance is taught. According to her, dance is a very creative art form—replete with creative expression and imagination. But the study of it is a discipline and requires a very rigorous approach in its teaching. A student of dance must undergo a rigorous training period, practice for hours on end, and rehearse individual steps and movements in isolation until perfect. It is only then that an entire choreography can be put together and performed flawlessly. The same approach would be true in the mastering and learning of other nonacademic disciplines such as vocal and instrumental music, gymnastics, painting, acting, and so forth. Any field that is nonacademic, that is creative and imaginative, still requires of its students relentless practice to perfect that skill. Thus it is reasonable to apply the same approach to the acquisition and mastery of academic skills. However, the strong focus on academic and formal learning in Indian schools does not seem to be detrimental to children's growth and development since, by and large, children in India are not socially and emotionally incompetent, and most do grow into adults who are stable and productive individuals with a keen sense of awareness of their responsibilities to their families and communities.

The next educational goal that was prioritized by the participants was teaching students to respect cultural and religious diversity in the classroom, school, and society.

Aim 4: Recognizing and Supporting Cultural and Religious Diversity

The recognition and celebration of diverse religious holidays is supported by India's Constitution, which takes a secular stance regarding the relationship between education and government. As a reminder to the reader, secular in the Indian context is understood as all religions being equally recognized and celebrated, and not the kind of separation between church and state as is the case in the United States. As a result most of the major religious holidays in India are observed and celebrated as national holidays. Each of the participants in my study shared that teaching students to acknowledge and celebrate all religious holidays was an important goal for them.

At the time that I visited schools for the purpose of this study, both *Christmas* and *Eid* (a Muslim holiday) had just passed, and all the classrooms I visited in New Delhi were still embellished with decorations and exhibits for both these holidays. Some of the other religious festivals and holidays that the teachers had celebrated with their students over the course of the school year had been the Hindu holidays *Diwali* and *Janamashtami*, and the Sikh holiday, which honored *Guru Nanak's Birthday* (Guru Nanak was one of the foremost leaders of Sikhism), amongst others. Charu recalled the celebration of *Christmas* in her school, and also described to me how this Christian holiday was incorporated in her daughter's curriculum in the higher grades:

> . . . all the religions are celebrated in all schools. Children in India like Christmas very much because it's such a jolly festival and they can participate in it. At school also we had a Christmas tree and they brought decorations . . . there was a special assembly where there was a Santa Claus. Then they had an entire scene put up on the stage where Jesus Christ was born . . . Yesterday my daughter had to write a little paragraph on Christmas in Hindi for her Hindi class. It was a little difficult to spell Jesus Christ and Bethlehem . . . But she knew the story and she could write quite a bit on her own.

Vasudha, the nursery class teacher, spoke of helping students learn to respect each other's cultures by participating in the celebrations of different holidays:

> I've got a Muslim girl and whenever there is *Id* [the Muslim holiday celebrated nation-wide] this Muslim girl will get *Sevain* [a sweet Vermicelli

pudding] from home, and she'll talk about her festival . . . Then for *Gurpurab* [the Sikh festival] we have a special assembly where we do the *Guru Granth Sahib ki pooja* [a Sikh prayer ceremony] and whatever they do, and then parents get *Kara Prashad* [blessed food] from home Then *Janashtami* and *Diwali* and *Ramayan* [three major Hindu holidays)—all that we have. Then we have Christmas celebrations. . . . for every festival there is a lot of festivity.

Vasudha highlighted the process by which children in the school celebrated the various holidays and how they all came together in the common celebration of national holidays, invoking a spirit of patriotism and unity: "whether the child is a Muslim or a Hindu or whatever religion or whatever culture that child might belong to—when we have Independence Day celebration then everybody is together . . . we're all united and we're all Indian . . . For Republic Day we're all together . . . ."

Malvika maintained a balanced perspective on the celebration of holidays and the teaching of culture in the curriculum. She noted that in the celebration of holidays it was important for teachers not to single out one particular religion as being more important than others, and they "shouldn't make the child very superstitious or very rigid . . . shouldn't make him stand on a higher pedestal than the other children from another religion."

Foreign conquests of India over a period of several centuries resulted in an on-going enrichment of the Indian culture due to the assimilation of diverse cultural elements. In general, this attitude of acceptance rather than stiff resistance to new ideas has kept the Indian culture pliable and thriving over a long period of time. Hinduism, in fact, is known to be the oldest continuing "religion" or culture in the world. Dave (1991) characterizes the Indian personality as one having an inherent love for a peaceful life and a natural aversion for aggressiveness, and is marked by nonviolence, cooperation, and tolerance. It is a historical fact that in her thousands of years of history, India has not been known to invade another country, and her freedom from colonial rule finally came from the nonviolence and civil disobedience movement led by Mahatma Gandhi. Acceptance of religious diversity is evident in many aspects of daily life in Indian society. An example is found in one of the nation's leading newspapers, which includes a daily feature called "Sacred Space." Each day, a topic or theme such as charity, humility, fear, war against evil, admitting a mistake, and so forth is highlighted by quoting texts and words from different philosophies and religions. Taking random examples, for instance, the newspaper edition of January 11, 2002, focused on the theme Love Your Enemy ("Sacred Space," 2002). Words from the following scriptures were quoted: *Dhammapada* (Buddhism), *Matthew* 5.43–45 (Christianity), *Vitaragastava* 14 (Jainism),

Asa-ki-Var, M.2, p. 474 of *Adi Granth* (Sikhism), *Yasna* 60.5 (Zoroastrianism), *Yuddha Kanda* 115 from *Ramayana* (Hinduism). The edition of January 17, 2005, focused on the theme of *Dharma* and included quotations on the importance of doing one's duty not only from the texts of *Dhammapada* and *Bhagvad Gita* but also from Dante's Il Convito ("Sacred Space," 2005).

Diversity in the form of multiple religions, racial and ethnic groups, and languages; multiple representations of gods; multiple realities of the Self; and multiple roles and responsibilities for the individual has shaped the social and secular nature of education in India. At the time of her independence, this secularism and recognition of diverse cultural elements was written into India's Constitution. In addition, a three-language formula adopted as part of the national school curriculum in India ensures that most school-going children in the country are bilingual or even trilingual. Thus, under constitutional law a form of multiculturalism and multilingualism is automatically incorporated in a school curriculum. This gets extended further as schools actively practice celebrations and cultural activities that draw upon different faiths and cultures.

India's openness to diversity is also influenced to some extent by the Hindu scriptures. The *Vedanta* explains the unity of the universe in terms of nonduality, *advaita*, which literally means "not two." It is this basic tenet of Indian philosophy that drives home the fact that diversity exists at a superficial level and is an illusion, *maya*. It implies that at a deeper, underlying level, all differences have a common source from which all life has originated. Based on my own observations and experiences, many of the school activities that celebrated diversity were focused on what multicultural educators in the United States would categorize as the superficial level of culture that includes tangible and visible expressions of foods, festivals, artifacts, costumes, and so forth. This to many Indians would not seem to be problematic because diversity is viewed and understood to occur at the superficial level. Of greater importance to them would be the deeper understanding and recognition that the underlying values across various cultures are similar rather than different.

This is not to say that conflicts arising from cultural differences do not take place amongst Indians. Tensions, quarrels, and feuds may occur frequently within and amongst families. Certainly many interreligious riots, mainly between Hindus and Muslims, have erupted in India, sometimes taking very violent forms. Mostly, however, instances of violence have been locally contained and not allowed to spread into a nation-wide situation. Friedman (2002) suggests that this could be due to the fact that centuries of Indian Muslims and Hindus living together in villages and towns has led to the sharing of communal institutions and a certain mixing of cultures and beliefs; further, the freedom of the press to report on such incidents can put pressure on government officials to take appropriate action as it happened a few years ago in the communal riots in Gujarat. Religious practices begin to have more and more elements that overlap over a period of time. The fodder that feeds the violence between Hindus and Muslims in India has more to do with India's political history than with religious intolerance. A deepening of this rift may be attributed to the time of India's independence when British colonial rule ended, but only after successfully partitioning the subcontinent into Hindu-dominated India and Muslim-dominated Pakistan. This geographical and religious division of the people of one country with one history heightened the animosity between Hindus and Muslims based simply on political and territorial competition. Prior to India's partition, Hindus and Muslims, just like Indian citizens from other religious backgrounds, had coexisted and had fought alongside each other against a common colonial enemy. The notion of diversity characterized within the Indian worldview is eloquently captured in the following words of Jawahar Lal Nehru, India's first prime minister (cited in Kakar, 1981):

> . . . though outwardly there was diversity and infinite variety among our people, everywhere there was that tremendous impress of oneness, which had held all of us together for ages past, whatever political fate or misfortune had befallen us . . . I was also fully aware of the diversities and divisions of Indian life, of classes, castes, religions, races, different degrees of cultural development. Yet I think that a country with a long cultural background and a common outlook on life develops a spirit that is peculiar to it and that is impressed on all its children, however much they may differ among themselves. (P. 14)

So far, my discussion has centered on the four primary goals of education from the perspectives of different educators in New Delhi; the goals include the teaching of values and attitudes, fostering intellectual development, teaching academic content, and supporting cultural and religious diversity. Although the roots of three of these educational aims can be traced to Ancient Indian philosophy and the Indian way of life, the fourth goal of academics and teaching to the test appears to have been shaped largely by the influences of British colonization.

Contextually Embedded Teaching Objectives

The sociocultural context of the study revealed that the values and attitudes teachers aimed to instill in their students were a way of life with people in

India and were reflected in their everyday behavior. For the purpose of this discussion, I have defined "values" as a way of life determined by a system of beliefs underlying the culture of a society. In this case, the culture is that of Indian society, and the system of beliefs is the "Indian-ness" that determines the way of life, or attitudes, in India and that distinguishes the Indian people. Many of these values are not related today to any one religious or ethnic origin but are found to exist in families at a deeper and wider level, crossing religious and linguistic differences and contributing to their Indian-ness. For instance, the importance attached to attitudes, such as hospitality toward guests, and values, such as respect for one's elders and teachers, is found to prevail among Indian families representing many different faiths and socioeconomic classes. These attitudes and values have their origins in an all-pervasive philosophy, which itself is not a religion but teaches a way of life. In addition, because this Indian philosophy, ancient though it is, is a lived philosophy and this way of life is actively practiced, most people living in India regardless of their religious or socioeconomic backgrounds begin to live aspects of it, thus constituting that "Indian-ness" I have mentioned several times.

Malvika, the first grade teacher, expressed it accurately when she maintained that although it was not explicit in the school's curriculum, the teaching of Indian values and beliefs was implicit in teachers' practice:

> In fact, the whole of culture is a part of your education. Like, see we don't even use this word [culture] in our curriculum but it's there . . . Whatever themes we choose to teach . . . are based on our culture. Let's say, a few months back we had a theme called Sharing and Caring, which is a part of our culture. This is what we've imbibed through the ages—sharing things and caring for others, love and care everywhere . . . whatever themes we've been trying to introduce have a very deep-rooted relation with education . . . the values we are talking about come from our culture.

Some of the more prominent values and attitudes that emerged during my conversations with the participants, namely respectful attitudes, social nature of the self, hospitality toward others, coexistence with others, and the influence and interdependence of the extended family, are discussed in greater detail in the following sections.

Respect for Elders

Respect for elders, be they parents, family members, teachers, friends or public leaders, is an important value that is actively practiced by Indians across socioeconomic classes and castes. Charu asserted this belief and explained how she reinforced it with her own daughters:

> I tell them . . . "have you ever seen me talking rudely to *Naani* [mother's mother]? *Never* [emphasis], even now. Even if I don't like anything then I'll tell her nicely. Have you ever seen me being rude to my father or to my mother?" But these days children will shout and scream and throw tantrums.

In her own second grade classroom, Charu promoted the teaching of respect toward parents and teachers by explaining to her students, " 'your parents will always think the best for you. So listen to them. Teachers are always doing what is best for you.' So this also they [the students] must incorporate, besides being sure of what they want [meaning besides the children making their own decisions]." Vasudha, too, spoke of several virtues and values that she wanted children to learn, including respect toward elders and all living things.

> There are so many values . . . loving and caring not just for humans but for animals, for every little thing that is around them—sharing with their friends, sharing with everyone around, then hospitality, honesty, socializing, justice, mercy, discipline . . . good manners towards their adults, respect for the elders . . . taking care of . . . all living things, living as well as non-living things—cleanliness.

It became clear from the several conversations I had with the participants that respect toward adults could take many forms such as exhibiting good manners, not being rude to them, being polite, volunteering for helping with errands and carrying heavy things when with someone older, not sitting down until every one older has sat down first, not eating until the elders have eaten first, and so forth. In general, it implied that age was venerated and that each individual whether adult or child should demonstrate extra care, affection, concern, and assistance to anyone older than themselves.

Malvika referred to values as "Indian values" and strongly felt that to counter the powerful influence of Western media on children in India these days we "first have to tell them that the Indian culture is a very strong one. We have to make them proud of our culture, and not feel that . . . it's only the others [other cultures] which are the best and ours is not." She described how she encouraged respect for a teacher in her classroom by telling her students, "whenever any teacher is asking you questions, you must stand up . . . have this much respect for your teacher when she's talking to you— you stand up . . . ." Teaching children to be respectful to adults does not necessarily detract from the importance of fostering self-confidence and initiative in the children. The two can coexist, and children may be taught that specific attitudes and values come into play in specific situations. It is a common misconception that this respectful attitude of Indian children toward adults makes them subdued and submissive in the company of adults. On the contrary, many are encouraged to voice their opinions and disagree with adults. But when in disagreement, they are expected to dis-agree in a respectful and polite manner rather than being rude or aggressive in their arguments. Children thus learn at a young age that they play mul-tiple roles, and while they must respect those who are older than them, they in turn will be respected by those younger. This give-and-take phenomenon works to ensure that no individual is permanently in a state of either authority or subordination. An individual is expected to be capable of exhibiting the qualities of a leader as well as a follower, as called upon. Thus the individual develops an adaptive and flexible attitude, knowing when to receive and when to give, when to lead and when to follow. Translating this into the early childhood school curriculum, it would imply that leadership qualities are not necessarily an important criteria for measuring socio-emotional development in children in an Indian classroom. This ties in with the perspective on self-concept, which is discussed next.

A Strong Concept of Self

Developing a healthy concept of the self was held to be very important by Malvika, and she repeatedly spoke of her aim to help children become self-confident and to develop a good sense of the self, or a strong self-concept. She felt that it was critical for teachers to help children:

> learn to have the feeling that we respect whatever we made . . . Have—sort of confidence in your hands, in your palms, in your fingers—yes, I can do any-thing in the world. That has to be incorporated at this stage. In Hindi we call it *apna haath Jagganath* [I have the power of my will and my own hands to accomplish what I want and shape my destiny]—that means your hand is everything, you can do anything in the world with your hand.

This is similar to the concept of *karma*, that life is not predestined but an outcome of the choices we make; both these attitudes are an accurate description of the Indian psyche. There is a difference, however, between Malvika's approach of developing a good sense of the self and the Western approach to building self-esteem, and this was reflected in my interviews with the teachers as well as in my classroom observations. It was evident in all the classrooms that I observed that rather than praising everything a child did or said the teachers were honest about their praise and offered it only if they felt the child had deserved it. Teachers and other adults were equally honest about matter-of-factly pointing out something that the child had not done, said, or written correctly. The result of this approach was that there was emphasis on helping children develop an accurate sense of self rather than on making them feel constantly good about themselves. It seems logical that the consequence of this would enable children to grow and develop knowing exactly what their strengths are without having a false or inflated sense of self-esteem.

During my observation of Malvika's classroom, both the class teachers walked around the room as the students worked on an English writing assignment, checking children's work as they passed: "This should be in capital letters. Why are all the letters small? The answer should be on the next line. It should be very neat" were reminders from the teachers as the children worked. The teachers encouraged the children to be attentive, finish their assignments on time, and not talk. Off and on, as the teacher passed by a child would inform her about his/her progress with enthusiasm. One girl said "Ma'am, I'm on my fourth sentence!" and the teacher responded by saying "That's very good." There was a lot of positive reinforcement of this kind, in spite of the frequent disciplinary instructions. All along, the teachers were leading the children forward in their work by gently and politely prodding and reminding them to go from one sentence to the other, "Now you should be at least on the third sentence. Jai, which number are you doing, still the first? Ankita, which number are you on?" Other children continued to walk up to the teachers for confirmation and affirmation of what they were doing, and the teachers gave honest feedback—they praised the work if the children had done it correctly, and explained otherwise if they had done something incorrectly in terms of wrong answers or spelling mistakes.

A similar approach was seen in Vasudha's Nursery classroom for four-year-olds as I observed the English Language lesson. The environment in both classrooms was comfortable in spite of the frequent directions and instructions. Teachers were comfortable and confident in offering both praise and criticism honestly and freely, and children seemed to receive the same with equal ease. Receiving criticism comfortably is determined to a large extent by the fact that the recipient is assured of the giver's love and affection. All the teachers I observed seemed to be aware of this and therefore made every effort to establish an affectionate relationship with the children in their classroom.

Hospitality toward Guests

Another value that is central to the Indian worldview and which the teachers appeared to emphasize in their classrooms is the importance given to guests and strangers—people who represent the "other." The notion of hospitality is pervasive and made a strong appearance during my meetings with the participants as well as in teachers' practice as evident in the classroom observations I made. During my observation of Charu's second grade classroom, I noticed how she reinforced this value when the opportunity came up during a story she was reading aloud to the class from the prescribed English Reader. In the story, a character named Mr. Ali visits the home of some children and is welcomed and invited in by their mother. Charu deliberately paused there in her reading and took the opportunity to reinforce the concept of hospitality and being welcoming to guests. "You should always greet people when they visit your house and offer them a glass of water or a cup of tea," she explained earnestly.

Hospitality toward guests stood out as a significant and important feature and was played out actively throughout the duration of my visit in Delhi—whether I was meeting with research participants in their own homes or at school, with former as well as new professional colleagues, or with my own family members and friends. Without fail a glass of water, tea, and snacks, if not a meal, was always offered to me persuasively. Having lived most of my life in India, in my experience hospitality was not just a characteristic of this one visit but a behavior that is exhibited by everyone in India at all times, and has been traditionally practiced in most Indian households, even by those families who live outside of India, elsewhere in the Indian Diaspora. There could be several reasons for this practice to have originated, and one interesting reason was given to me by a friend who is a native of the desert state of Rajasthan. She said because India has such hot summers with a scarcity of water, offering a glass of water is a considerate and thoughtful gesture when someone visits you. But a stronger and deeper explanation is possibly found in the ancient scriptures where it is clearly written, "*atitih devo bhavah*," which implies that God can choose to visit your home in the guise of a guest. Thus across the various socioeconomic classes and castes in Indian society it has been ingrained that a guest should be welcomed with the utmost respect and hospitality, and this belief is practiced actively and widely. Even today, extending hospitality toward a guest or a stranger is accorded high priority in the scheme of duty.

When I visited Malvika's house for my first interview with her, Malvika's mother served us three times during the course of the interview. First, as soon as I was seated, she brought me a glass of cold water. Then a tray laden with snacks and sweets such as *gaajar ka halwa* (a halwa made out of grated carrots cooked in milk and sugar) and *barfi* (a sweet made out of milk and sugar) was placed in front of me. Finally, she handed me a cup of hot, freshly brewed tea at the end of the interview before sitting down to join us. At the end of the interview session her brother, too, joined us and

I conversed with the family as I sipped my tea. They were polite, respectful, attentive, and very hospitable. As I thanked them and prepared to leave, Malvika and her brother offered to walk me down the street to the *riksha* stand (a *riksha* is a three-wheeled scooter that is partially enclosed and is used extensively as a form of public transportation, much like a cab). On the way out, we met her father returning from outdoors. Malvika and her brother stopped to greet him and introduced me to him before proceeding to make our way outside. I was struck by the gracious hospitality the entire family had extended toward me—a stranger visiting their home for the first time.

I experienced the same degree of hospitality when I visited Vasudha's house where I met her for two of the three interviews I conducted with her. On the first visit, Vasudha's mother-in-law offered me a glass of water as soon as I was seated, and then within 15 minutes of starting the interview, Vasudha's husband brought in a tray with a kettle of hot tea, hot *samosas* (a popular savory snack stuffed with spiced potatoes), and a dish of cheesy pasta that Vasudha herself had baked especially for my visit. During the second interview two weeks later, I was served a scrumptious snack consisting of meat patties, *samosas*, and tea. These are a few examples of the kind of attention, care, and gracious hospitality that I experienced not only during all the professional visits, but even during personal gatherings with my family and friends. It merely reinforced what I have always known: that hospitality is an integral part of the Indian way of life.

The Phenomenon of Coexistence

The concept of multiple realities is strikingly evident in Indian society. During the course of my fieldwork, I noted that boundaries between professional and personal were less defined, and professional work in the form of my research interviews seemed to merge well with personal interactions within the home environment of the participants. The concept of closed doors and privacy did not seem to be an issue at all and participants were comfortable speaking to me even when family members were in and out of the same room. This occurred several times with each participant. During their respective interviews, Malvika's mother and brother were in the same room listening to our conversation. So were Vasudha's husband and daughter, Nagma's mother and husband, Ashwini's coworker, and so forth. The ability to switch from one situation to another with ease and comfort was noteworthy. Participants could be simultaneously attentive to different people who demanded their attention. Let me describe part of my third interview with Charu, which took place in her home. As we spoke, the phone rang and interrupted our conversation. Charu took the call and chatted with the caller for a few minutes, calling her "aunty." Apparently Charu's mother had been unwell and was sleeping in the next room. Charu was attending to her and taking care of her until she recovered. Charu did not apologize after the phone call had ended. Neither of us had thought in terms of an apology as interruptions were not viewed as interruptions, merely events that happened alongside others. Coexistence. All could occur and exist simultaneously and be given equal importance. In the Indian mind set, she was not detracting from the importance of our conversation by taking the phone call from her family and speaking for a few minutes. It was not an insult in her mind and it certainly was not an offense in mine. I was content to wait for a few minutes till I had her attention again.

The layers of culture that make up the Indian personality appeared to exist with several distinct but overlapping elements of Indian tradition and Western styles. Adults and children switched rapidly back and forth between Hindi and English, often speaking in Hindlish; jeans, *saris*, and *salwar kameez* were worn with equal grace and comfort by women; cheese pasta and spicy *samosas* were served simultaneously on the same tray; decorations and children's artwork for both Christmas and Eid were seen side by side in classrooms (both holidays had occurred in the same month); both traditional elements and progressive ideas were apparent in the languages used, in the décor of homes, in the names of people, in the school's curriculum, and even in the news and media.

I visited Charu's home one afternoon in December for our first interview. Neighborhoods in India are known as colonies and the neighborhood she lived in was a clean and quiet gated colony since it was exclusively for government officials. Inside the colony were double-storied houses with front and back lawns. I located her house easily and rang the doorbell. A young girl, about twelve years old, opened the door. She greeted me and asked me to "please come in and have a seat." I introduced myself and asked for Charu. The young girl informed me that "Mummy has just gone out and will be back in a few moments." I took a seat in the enclosed front porch amidst a couple of easy chairs and a coffee table. A *diwan* (a flat couch with no back rest or handrests) was placed along one wall and had a *harmonium* (an accordion-like traditional Indian musical instrument) resting on it. The porch was decorated with Indian fabric, wickerwork, and sculptures. A piece of driftwood adorned one corner of the porch. A doorway was covered with straw matting trimmed with block-printed fabric. The glass panes of the door were decorated with Christmas ornaments and stickers—bells, trees, and lots of santa clauses. The porch overlooked the front lawn. The sun had moved across to the back of the house and there was a slight breeze blowing on that wintry afternoon. As I waited for Charu's return, I soaked in the silence, which was interrupted only by the calling of birds in the trees. A few minutes later, a car pulled up into the driveway that was half hidden by the hedge. A woman in her early thirties and a young girl of fifteen walked into the porch. The woman was Charu. She greeted me and apologized for having to run out for an errand. Apparently one of her friends had wanted her to pick up some movie tickets from the theater close to Charu's house and she had willingly and happily obliged. Charu offered me a cup of tea and I deferred but agreed to have it after the interview was over. Charu was wearing blue jeans, a sweater, and sneakers. Her long black waist length hair was left loose. We settled down to get started with the interview.

My first interview with Vasudha, on the other hand, was a complete contrast and took place in a very different setting. Vasudha lived in a vast colony that was congested and heavily populated with crowded living areas. I found her house with some difficulty after having to ask around. I realized that Vasudha's directions on how to get to her house had been accurate but the houses and buildings in that section of the colony were numbered in some illogical manner. Outside her narrow double-storied house was a *shamiaana* (a large colorful canvas tent-like enclosure, which is erected for outdoor events) and a wedding was going on. Music blared on loud speakers and there were many people gathered on the street and seated in the *shamiaana*. As Vasudha and I began to talk, we were interrupted by her three-year-old daughter who was asked by her mother to leave and go to an inner room. Vasudha made this request in a firm but gentle manner. Then, sensing that there might be further interruptions, Vasudha asked me if we could move our meeting to the second floor of the house, which was her floor. I agreed, and in order to do that we had to walk through her parents-in-laws' bedroom where they were sitting watching television. We walked up two flights of narrow stairs to her suite of rooms. There I met her husband and some friends who were visiting them. The two of us sat in the outer living room to complete the interview while her husband entertained their friends in the bedroom inside. I was struck with the contrast of Indian and Western in Vasudha's house too. Vasudha, in her jeans and sneakers, was a contrasting image to many of the ideas, words, and behaviors in the house. The house was full of things—people, furniture, stairs, walls, curtains, swings in doorways—in no particular order or theme, all unrelated but existing together. There was a bulletin board on one wall that Vasudha had decorated for Christmas with green and white balloons. It was a sudden reminder to me that it was indeed Christmas Day! Christmas was being recognized and celebrated in the form of decorations in the home of this Hindu family but it was clearly not a religious holiday for them. Life continued and it was just fine to schedule interviews and complete chores on this holiday. It was a house inhabited by a joint family: Vasudha's parents-in-law; Vasudha along with her husband and daughter; and Vasudha's husband's older brother and his wife and two daughters. It was a home where three families lived together, displaying their various coexisting personalities in a mingled but nice sort of way.

This phenomenon of how life in India is an intimate overlap of contrasts was supported by several other examples, which illustrated how elements and ideas from different time periods in history, the different cultures, languages, beliefs, clothes, celebrations, styles, and so forth existed simultaneously and that people in India borrowed freely from different traditions to make it all a part of their natural, every day world.

Joint Families, Extended Families: Roles and Responsibilities

Family connections are strong and intergenerational coexistence is also evident as with Vasudha, who lived in a typical joint family system. In an extended family structure, members may not live under the same roof but have frequent interactions and are readily available to help out whenever needed. Charu recently lost her father and now felt completely responsible for caring for her mother who also lived in New Delhi. On the day of my second interview with her, she told me how she had taken her mother out shopping for gifts that the latter was going to carry for her brother's family in another town. Nagma, one of the teacher educators, lived with her husband and her nonagenarian mother. On the day I interviewed her in her home, our conversation was interspersed with a conversation that Nagma's husband was having with her mother as they had tea seated at the dining table a few feet away. It didn't seem to bother anyone except, perhaps, the tape recording. It was just another instance of coexistence: young and old, personal and professional, suggesting coexistence to be the very texture of Indian society.

I was aware of this phenomenon in the families of other teachers and educators also whom I met during the course of the fieldwork. Alpana, a Pre-K teacher, recently lost her father to cancer. She had taken a month off from work to help her mother care for him, and her sister had come all the way from North America to help. Another teacher, Chandrika, was a cancer survivor who also cared and provided for her divorced daughter and three-year-old grandson. Zubeida, a retired teacher educator, was also a cancer survivor and had a daughter who was divorced with two young children. Zubeida still visited them and supported them in any way she could. On a personal note, my parents who are both practicing physicians find the time and energy to care and look after my father's father who lives with them and is now 98 years old. To rephrase the previous sentence, finding the time is not an option. With Indians, the internalization of the concept of *dharma*—duty and moral responsibility—is a reality. Time must first be made available for family responsibilities and then one tries to fit in the social and recreational activities for oneself. The "other" is prioritized over the "self." But somewhere down the line almost always the responsibilities start to feel less like duty and more like love and caring. My own brother and his family, after having lived in the United States for 15 years are now, of their own volition, moving back to India to live with my parents and grandfather so that the latter too can be cared for in their old age.

When during the course of one of our conversations I asked Vasudha to speak about extended family relationships, she emphasized what she called "the strong family bonds in India." Vasudha shared that it was because "our lives [the lives of Indian parents] only revolve around our kids . . . Life abroad is quite hectic, full of machines, and everybody is so intent making their own careers. Even here people want to make their own careers but not at the expense of family life." She observed that in India it was not only the parents who were so attentive and tuned into their children, but also children, who when they grow into adults mirror the same protective attitudes of care and concern toward their aging parents. She illustrated by explaining, ". . . in my own family I see my own brother all the time bothered [feeling concerned] about my parents, and here my husband and brother-in-law care about their parents. They really care about their parents—all the time. Whenever we're going out they make sure the [telephone] numbers are left with them . . . in case of any emergency."

While I was talking to Chandrika, she told me the moving story about her struggle with cancer. She had no support from her husband, neither financially nor emotionally. When she had to undergo radical surgery, it was her school that came to her assistance. The principal gave her a huge monetary loan, and relieved two teachers to be with her at the hospital during the time of the surgery. Chandrika subsequently started a cancer support group, one of the first in Delhi. She shared with me that she valued the strong commitment that her school had demonstrated, and that she would rather have this support than fight for more independence or a stronger voice as a teacher in the school. Loyalty to the school and the profession was born out of a symbiotic relationship between the teacher and the school, not by being a more independent teacher. This was a good example of the kind of community building and extended family feelings that are fostered in many schools and communities in India.

Translating Educational Aims with Respect to *Dharma* and *Karma*

The critical curricular issue of what is worth teaching and who makes this decision emerged in an interesting manner in the light of the findings of this study when viewed in the context of early childhood teacher preparation and practice. Of the four primary educational aims discussed above, three of them, namely teaching of values, intellectual development, and respect for diversity, appeared to be strongly influenced by Hindu philosophy.

Not only is the concept of *dharma* and *karma* seen to be central to the educational objectives described above, but it also appears as fundamental to concepts such as social development, emotional development, and moral development, terminology that is couched in the discourse of Western early childhood education. American and European progressive traditions in early childhood education do consider the "whole child," but view the development of social, emotional, cognitive, and physical skills as individual though overlapping domains, with specific learning and psychological theories defining the nature of development within each of the domains. In Indian philosophy also, the growth of the individual is viewed holistically, with the concept of *dharma* and *karma* implying the overall growth and learning of individuals in relation to each other, and in the context of their duties and responsibilities to each other in order to maintain harmony in society. The child within the Indian worldview is considered both as a social being and as a unique individual (Viruru, 2001). The concept of *karma* emphasizes the individual in terms of the choices one makes, one's actions, and the consequences of these actions, whereas the concept of *dharma* emphasizes the relationships the individual has with family and society. This approach considers a child's growth and development as a complex, three-dimensional whole, without different theories governing the development in individual developmental domains. The findings of this study also indicated that developing an overall, all-rounded personality in the children was an important objective for the teachers. The adult–child continuity that is discussed in more detail in a later chapter, results in upholding similar behavioral expectations for both adults and school aged children. Some of these expectations are listed in figure 4.2 in which I have presented a conceptualization of how *dharma* and *karma* define the connections between education and daily life in India.

Certain patterns were enmeshed in the data and emerged as definite themes during data analysis. These themes identified by the values and attitudes rooted in the Indian worldview can be placed into the categories that I have delineated in figure 4.2. But in spite of having already discussed these

Dharma and *Karma* expressed in themes that emerged from data analysis

As manifested in daily living	Carrying out one's duties toward each other
	Being aware of one's responsibilities toward family and society
	Showing respect toward elders
	Showing respect toward teachers and institutions
	Being balanced in one's emotions and behaviors
	Being aware of the environment and its protection
	Living multiple roles with multiple responsibilities
	Realizing materialism as *maya*—an illusion
	Attaching much importance to knowledge and learning
	Seeing self as part of a group as well as an individual
	Recognizing each as a part of the Divine, which connects all
	Having clarity of thought to make the right choices
	Being mindful of the consequences of one's actions
As translated into an educational context	Maximum importance to intellectual development
	Moral development: strong sense of right and wrong
	Social development: sense of duty, respect, responsibility
	Emotional development: maintaining one's emotional balance
	Fostering diversity and recognizing the one-ness of all
	Overall development of the child's personality, character
	Completion of tasks and assignments valued over verbal/leadership skills.
	Little importance given to sensory development

Figure 4.2   *Dharma* and *Karma* in relation to daily life and education in India

categories separately, I feel the need to reexplore these issues from the overlapping perspectives of respect, duty, and multiple roles and identities and in special relation to language.

Respect is a basic attitude of the Indian personality and is important to the Indian psyche. In the *Bhagvad Gita*, Lord Krishna proclaims that God resides in the heart of every individual, with the implication being that every individual needs to be respected and honored. Another saying is *vasudaiva kutumbakam* (the whole universe is my family), and by saying these words the Indian mind elevates the concept of brotherhood to a lofty level (Dave, 1991). The daily common greeting in Hindi between any two individuals is *namaste*, which literally means, "I respect the divine in you." *Namaste* is used as a universal greeting for hello, goodbye, good morning, goodnight, and so forth. Other Indian languages derived from Sanskrit use some variation of the same word such as *namaskaram, namaskar*, and so forth. Most Indians, regardless of their cultural background, do recognize the belief that there is a part of the Divine Energy in every human being, which is why being respectful to the "other" is so important in order to be of good character. Another linguistic indicator of how this concept is built into the values and practices of the people in India is in the use of the pronoun "you." In various Indian languages, the word "you" takes different forms depending on the respect or affection provided to the person being addressed. For instance, in Hindi, all elders, teachers, guests, and so forth are addressed by using "*aap*," which is a respectful term. Friends and peers may be addressed by using "*tum*," which indicates a more casual and friendly relationship. Younger people may be addressed by the very informal and affectionate term "*tu*." Children are taught to respect their elders by not only their families and their teachers, but also through the use of daily language. It becomes the duty of the adult to reinforce in children the understanding for the right values for a good life. This correlates back to Vygotsky's idea that language has a tremendous influence on children's construction of meaning and is a subsequent influence on social attitudes and behaviors as discussed in chapter 1.

Another phenomenon that is closely related to the concepts of both *dharma* and the usage of language is the existence of kinship patterns. People live multiple realities from the multiple kinship roles they are born or married into. Each position in the kinship web has a specific title. For instance, in the Hindi language, which is most commonly spoken in North India, the same woman could be a *chaachi* (father's younger brother's wife), *tai* (father's older brother's wife), *maami* (mother's brother's wife), *bua* (father's sister), *maasi* (mother's sister), *didi* (an older sister), *naani* (mother's mother), *daadi* (father's mother), and so forth. The titles would be different in the different Indian languages but each role clearly comes with a different identity and a different set of privileges and rights. What is important to recognize here is the underlying understanding that with specific positions of power are attached specific responsibilities and duties that must be carried out. This provides every individual his or her turn at being in a relatively higher or lower position in the family hierarchy during a life cycle. Due to the extended family system, children are known to experience multiple mothering from all the women in the family. The oldest woman in the family is seen to have the authority and respect to make decisions for all children in a family. Children themselves may be multilingual, and learn to negotiate multiple realities early in life knowing which language to use in a specific context.

Intergenerational living was seen to be common in India and many actions in daily living fall under the umbrella of "duty"—buying gifts for family and friends, inviting friends and family over for a meal, visiting and making social calls. Even within the school environment I observed a hierarchy that was carefully adhered to: the classroom teachers followed the headmistress's instructions, who followed the principal's instructions, who, in turn, had to defer to the requests of the chairperson of the school board. With each role the individual assumed a different identity, a different code of behavior, and different mannerisms. It was fascinating to watch all these different individuals move effortlessly between their multiple roles and multiple identities—as women, mothers, daughters, daughters-in-law, wives, teachers, friends, employers, supervisors—with much ease and grace. It was very clear that an individual did not have one identity, nor was there an expectation to present one identity. In fact, multiple identities is the norm and expectation in India. This puts an interesting twist on the phrase so commonly used in the West "be yourself," because truthfully, each individual in India seemed to possess several social selves.

It is also interesting to note that Indians have a greater loyalty to their private worlds comprising family and close friends, than to their public world comprising work and professional colleagues. One outcome of this is that, in general, the concept of group work and collaborative projects does not find a prominent place in the school curriculum and workplace in India. School projects and assignments are individually done and evaluated. But the other outcome of this loyalty to their private worlds is that regardless of socioeconomic class, there is a huge degree of kinship and team spirit within families and communities, and members go through the ups and downs of life supporting each other. I observed during the course of the study that *group work* was not mentioned as a goal of education by any of the participants, although learning to *live together* and get along with each other was an important and frequently mentioned goal. There is also the correlation between this loyalty to one's private world and to the understanding of charity. Donating to charitable institutions is not a common practice in India. People in general are more comfortable and willing to help family members and friends who are vulnerable and in need, and it is very common to see families taking in relatives who need to be cared for.

It was clear from the conversations I had and from my observations that the values and attitudes that were prioritized by educators participating in my study and which teachers aimed to reinforce with their students were indeed a way of life with people in India, reflected widely and commonly in their everyday behaviors and life styles. In chapters 5 and 6, I turn to cultural constructs and examine how the current image of the teacher and child in India have been socially and culturally shaped by historical and contemporary influences.

Chapter 5

Image, Role, and Responsibilities of the Early Childhood Teacher in India

Based upon conversations with the participants, the questionnaires completed by them, and my classroom observations of three of the teachers, it was clear that there was a fairly strong correspondence between the traditional concept of the teacher in India and the current image and role of the early childhood teacher. This correspondence influenced the teacher's relationship with her students as well as the criteria that were used by school principals for hiring teachers. A description of my first visit to Malvika's first grade classroom will enable the reader to better contextualize the subsequent discussion on the image and role of early childhood teachers as commonly seen in private urban schools in India.

I reached The School at the appointed time in the afternoon, and after meeting the headmistress I headed toward Malvika's classroom. There was plenty of artwork displayed on the walls and the physical space in the room was taken up by tables and chairs. The students were attending a dance class at that moment but were due to return shortly. When they did enter the classroom they went immediately to their assigned seats. There was a total of 40 students. As they sat down, the two coteachers, Malvika and Jaspreet, asked them to get ready for the English lesson. Copies (notebooks) were taken out and opened. "How do you spell Monday? Only fourth and fifth rows answer," the teacher said. I gathered that the students sat in rows that were numbered one through six and only the students in rows four and five were being asked to answer that specific question. They spelt out the word and recited the date on cue—"Seven-one-zero two." Their response was indicative of the practice of writing the date before the month in India.

The work assigned for this class session had already been written on the chalkboard in prior preparation for the class. It was a quiz, or a competition. The students were divided into teams according to the rows in which they were seated. There were six teams comprising rows one through six. I noticed hints (clues) to the quiz in the form of a word list that had been earlier written by the teachers on one side of the board. Jaspreet asked a question—"Who will read the first word? Raise your hands if you want to read." She then called on students by name to read out one word each. A deliberate emphasis on right pronunciation was made as the teacher asked "do you say 'Munkey' or 'Monkey'? Is it 'Dunkey' or 'Donkey'?" Students giggled at the wrong pronunciations but the point had been made. Just then a child entered the classroom to convey a message to Jaspreet, who was leading the class. It was a message from the office and she had to leave the classroom for a few minutes. Malvika took over the lead immediately and continued the lesson without a break. Another question was asked. Several hands went up and many voices pleaded to provide the answer. Malvika announced that only the quietest row would get to answer. Most of the 40 students appeared to be attentively listening or writing. In the last row close to where I sat, two students talked amongst themselves for a few minutes and then returned their attention and focus to the lesson. As a nonparticipant observer, I sat in the last row next to a girl whose name, as it appeared on her notebook, was Hazel. I was struck by the Western name and noted that she had a Hindu last name. Hazel was wearing glasses and her notebook lay open to a page that had been checked by the teacher who had given her a grade of "10/11, Good." She raised her hand to respond to many questions. Whenever a question was asked several hands would go up to answer it. The students were attentive and eager to answer, and the high energy level in the room was very tangible to me.

After the quiz was over, students were asked to begin the follow-up written activity that involved writing down the quiz in their notebooks. Jaspreet had returned to the classroom by now. Both teachers provided explicit directions while the students were writing. The most used phrases were: "stay in your own seat," "please don't talk amongst yourselves," "raise your hand if you want to answer," "when you write be sure to put the sentence on one line and the answer on the next line," "write your work neatly and carefully, that's very good." Amidst all this, Malvika began to return some graded work sheets that had been given to those students who had been absent for one of the previous assignments. Both teachers continued to walk around the classroom constantly, checking students' work as they passed. "This should be in capital letters. Why are all the letters small? The answer should be on the next line. It should very very neat" were other reminders from the teachers to the students as they worked. "Excuse me,

Ma'am!" one girl said when she was unable to see the blackboard behind the teacher, and Jaspreet moved out of the way.

The bell rang to signal that a half-hour period was over. Students continued with their work as this happened to be a double period for English. I noticed that almost all students focused on their work even though some were talking to neighbors as they worked. Both teachers encouraged them to finish their assignments in a timely manner and not spend too much time on conversation. After a while, when some students had finished their work and completed all six sentences, the teachers asked them to draw the animals they had written about in their notebooks. This gave them extra work to keep busy with while other students were still completing their sentences. Several students walked up to the teachers for confirmation and affirmation of what they had been working on, and the teachers gave them honest feedback. They praised the work if the student had done it correctly, and explained otherwise if it had to be rewritten in the case of wrong answers or spelling mistakes. The overall classroom environment reflected a sense of busy-ness, whispering, and talking. But there was no confusion or chaos or the deafening sound one might expect in a room filled with 40 six-year-olds. It sounded like a loud but busy hum. Students were encouraged to stay in their seats, but were not admonished if they came up to the teachers or the black board for clarifications and questions. This description of a class in progress would be quite typical at an elementary grade level in most private schools in New Delhi.

The *Guru*: Image of the Teacher in Ancient India

The importance that has been traditionally attached to a teacher in India may be compared and contrasted to the importance that is accorded to an institution or one's Alma Mater in the United States. Traditionally, teachers in India have been viewed in the image of the *guru*, a highly respected teacher who would educate and care for a group of students in a school in Ancient India. The *guru* thus became the prototype for subsequent teachers in India. The word *guru* implies one who leads us from darkness to light, or in other words, from ignorance to knowledge. Since knowledge and intellect were so highly regarded in the ancient texts, the teacher whose duty it was to facilitate the development of the intellect was also placed in a very respectable and important position. One of the Sanskrit *shlokas*, or verses, that is often a part of many prayer ceremonies is a salutation to the teacher: *gurur Brahma, gurur Vishnu, guru devo Mahesvara, gurur sakshaat param Brahma tasmai shree guruve namaha* (the *guru* is none other than *Brahma*, the creator; *Vishnu*, the preserver; and *Mahesh*, the destroyer. He is in the image of the Supreme *Brahmaan*. To such a *guru* I offer my salutation).

In many ways, the *guru* was more revered and respected than one's own parents. To parents we owe our physical birth, but to our *guru* we owe our intellectual and spiritual regeneration (Altekar, 1965). The teacher was not only highly venerated but was expected to possess certain qualities that would make him worthy of that respect and veneration. The *guru* was one who would without any reservations place at the disposal of his students the essence of his knowledge and experience. A comprehensive and detailed discussion on the qualities that the *guru* was expected to possess, his duties and responsibilities, the high code of the teaching profession, the fee of the teacher, teacher–student interactions, disciplinary codes, and other educational issues of Ancient India may be found in the work of Altekar (1965). Briefly, the teacher was to be of high character, patient, impartial, and fair in the treatment of his students; well grounded in his own branch of knowledge; and continue to read and learn throughout life. Besides extensive scholarship, the teacher was expected to be fluent and adept in delivery and teaching, witty, possess good presence of mind, and, above all, inspire students through his piety, character, and scholarship. One of the vows the *guru* had to take as a teacher was that he would teach everything he knew to his students without withholding any knowledge in the fear that one of the pupils might outshine him as a teacher one day. It was expected that teachers would not take a fee from their students to impart knowledge and learning to them but would be satisfied with whatever *guru dakshina* (gift or honorarium) their student offered them but only upon the completion of their studies. Conversely, it was the duty of each student to offer his teacher *guru dakshina*. Records indicate that teachers held different ranks and were called *guru, acharya*, or *upadhyaya* depending on their rank and specific roles. The teacher had complete freedom in designing lessons and the curriculum for his students given that he was the only one teaching and governing his *ashram*. Healthy competition existed between teachers, and the number of students in the *ashram* was indicative of the success and quality of the *guru*'s practice. It is a fact that most teachers in Ancient India, as in other ancient civilizations such as Greece and China, were men. But records indicate the existence of women known as *gurumata* (guru mother), and it is possible that certain *ashrams*, or places of learning, were overseen by women if the students comprised girls and very young boys since coeducation was known to be prevalent in ancient forest schools (Achyuthan, 1974).

Although I have provided a description of the teacher in India during ancient times, a similar degree of veneration and respect continued to be accorded to the teacher in the ensuing centuries throughout the course of

India's history. Even today, the importance given to teachers is implied in the general understanding and recognition that students' learning is influenced more by committed and competent teachers and is less dependent on buildings, equipment, and material resources. A traditional view about the image of the teacher in India even today emerges most clearly from conversations with the practitioners—the teachers themselves and the principals of the schools. The attitude of being respectful toward teachers and educational institutions is grounded in the fact that in India, wisdom, learning, and the arts are personified by Goddess *Saraswati*. When I visited another prominent private school in the city, I had to wait in the reception area for a few minutes before I could meet with the principal in her office. In the hallway outside the reception room was a massive, floor-to-ceiling sized carving of *Saraswati* against one wall. What was interesting was that almost everyone—students, teachers, and parents—stopped as they passed that way and paid their respects to the Goddess by bowing their heads, folding their hands in a *Namaste*, or by touching her feet. It was deeply symbolic of how important the pursuit of knowledge and learning has traditionally been to the Indian psyche. Mehta (1997) writes about the prolific reading culture that is found in urban India, and despite widespread illiteracy the ability to read is treated with a sense of awe and respect. She describes a common scene in Mumbai where booksellers cajole and beg you to read by jumping onto your moving train, clinging to the bars of the window with one hand and pushing a basket overflowing with books toward you with the other hand. According to a 2005 BBC report based on a worldwide survey, the maximum time spent on reading was by Indians with an average 10.5 hours per week ("India's world's biggest readers . . ."). The survey also indicated that Indians spend the least time viewing television. Books in India are considered to be receptacles of knowledge and tools for sharpening the intellect, and to disrespect a book or any paper with writing on it by touching it with one's feet is an unthinkable disgrace. It goes without saying that any disrespect shown toward teachers, educational institutions, and even books is considered to be disrespectful of Goddess *Saraswati*. Charu voiced her dismay over the fact that many schoolgoing children can be disrespectful: "It's really bad, you know, talking against the system, talking against the teachers, talking against the school. I really feel bad."

Who Is a "Good" Teacher?

During one of my conversations with Shakti, the school principal, we spoke about what made a teacher a good teacher. Shakti emphasized that for her love was the biggest thing and a good teacher was one who, first and foremost, not only loves the children but loves herself for this is what gives her confidence. Only then is she able to stay happy, healthy, cheerful, devoted, and keen on learning. Shakti shared that during the teacher hiring process and in her evaluation of teachers, she appreciated teachers with values—not big things but small virtues such as consistency, hard work, someone who interacts with the children and who sees herself as a learner. However, the most important point that Shakti made was that a teacher could not teach values to students if she herself did not possess those values. When I asked her how this could be seen in the context of a teacher education program, Shakti replied: "We can't prepare them. They have to already have those qualities ... You must possess it yourself ... from culture, upbringing, realizations. You know, it can't just be words. Every teacher who imparts value education must possess the qualities."

Antara, the elementary school headmistress, described other attributes that a good teacher should possess, namely to be good in her communication skills, be presentable and creative in her ideas, to have a sensitive attitude toward children, and to be a patient listener. She emphasized the tremendous influence a teacher has on children's growth and learning:

> I have always felt that it depends on the teacher to develop the overall personality of the child ... A teacher greatly influences a child so she is the one who is responsible for the overall development of the child right from physical to emotional to cognitive to educational aspects of the child. In my teaching experience I have always seen that a child of school age is more influenced by teachers and his peer group than the parents. A teacher has a major role to play in shaping a child's personality.

This influence is the result of the recognition and respect that is generally given to a teacher by children, and more so by the families of the children.

Recognizing the teacher's influence over the children in her classroom is what shaped the practice of the three teachers Charu, Vasudha, and Malvika. Charu acknowledged that she had a responsibility toward the children in teaching them not only academics but also about values and attitudes because she knew that children in general listen more carefully to what the teacher tells them than they do to their own parents. In her conversation, she clarified her meaning by saying, "Subconsciously, parents make more of an impact. But in their conscious being, I think, they learn more from the teacher or they are willing to [learn] more from the teacher ... I can teach them [the students] but not my own children. Your own kids never listen to you. They never do what you say whereas children in school do." However, Charu also made it quite clear that a teacher cannot exercise her influence in an oppressive or dominating manner: " ... the teacher must try and build a bond with the children. That is very important. Like we do with our children at home. They must feel close to you . . . That is very very [emphasis] important, much more than anything else. But once they view you as a friend, or as someone they like, then whatever you say makes more of a difference." Because of the strong influence a teacher has on the children, Charu felt that was all the more reason why a teacher must be careful about how she presented herself to her students:

> . . . how she presents herself is very very [emphasis] important—her body language, for one, is very very [emphasis] important. You might not realize it but while you're there with them for those four-five hours they're watching you—every minute, every second . . . How you talk to them, or if another teacher comes into the room or an ayah or peon, how you react to them . . . So a teacher's behavior, her body language, how she responds to any situation, they're watching you and it has a major impact.

Malvika, too, echoed these views and described one particular Physical Education teacher in the school that she attended when she was a child, whom every one loved and respected so much that "students used to touch his feet because he was very caring and loving and he used to bless them every time they used to touch his feet." The actual act of bending down in front of someone who is older was probably done to make it more convenient for the older person to physically be able to place a hand on the head of the younger person who was respectfully asking to be blessed by an elder or a person in authority and was not at all meant to be demeaning. Symbolically, it is an opportunity for the latter to confer love, blessings, and good wishes upon the younger person. Let me illustrate by describing an incident that occurred at the end of my interview session with Nagma. As I was preparing to leave, Nagma introduced me to her mother who was sitting in a sofa chair, reading a newspaper in the same room where we had been interviewing. Her mother asked me in fluent English where I was doing my Ph.D. and I told her that I was studying in America. Then she asked me why I was in India and I explained to her that I was there for my data collection. She then smiled, raised her hand to bless me, and wished me good luck. I felt something very profound and special upon receiving the good wishes and blessings of a ninety-five-year-old woman.

The Responsibility of Teaching: A "Sacred" Profession

With age and authority comes a responsibility that cannot be underscored enough. The Indian system of social and familial hierarchy is not a power game in one direction but rather has a democratic two-way dynamic that may not be quite apparent to, or understood by, others outside of the Indian culture. Malvika made an excellent point of illustrating the nature of hierarchy as understood within the Indian worldview. With every position of authority or privilege comes a certain responsibility and sense of duty that is expected to be carried through by the individual concerned. In response to my questions about the image of the teacher, Malvika answered very earnestly and sincerely, offering a lucid description of how a teacher was viewed:

> Well, next to God! . . . especially I think for the younger ones, it's really next to God. Not exactly to worship or to pray to, but to follow . . . and imbibe whatever you can imbibe from the teacher . . . More than their Gods, they [the children] follow you [the teacher] . . . apart from imparting values, she [the teacher] has many other roles to play. She is almost like a mother for five-six hours [in the classroom] . . . To tell them all the good things in life, to make them aware . . . To make them a little more rational . . . she is a nurturer . . . a counselor . . . a facilitator . . .

In continuing her description of the privilege and respect a teacher is owed, Malvika underscored the responsibilities that come with the importance and respect given to the role of a teacher:

> But for that matter, the teacher should be an ideal one. If I don't have certain values that I want the children to pick up then where exactly are they going to pick them up from? . . . they have their own home environment where they pick up certain values but we've got our own share of duties that we should do—as a part of our job and as a part of being the teacher—it's such a sacred thing to be—such a sacred profession. So they [teachers] need to be . . . ideally suited to the profession . . . I feel there is a strong connection [between a teacher's professional and personal values] . . . And the kind of behavior, the code of conduct that the students show actually reflects what the teacher has taught them.

Vasudha also shared her views, which corresponded with those of her colleagues. She expressed a teacher's role as being important and pivotal because a child listens more to his teacher than his parents. She felt that a child's respect for a teacher could only grow out of the loving bond that she is able to establish with her students. Consequently, with this kind of a relationship the teacher is the one who can bridge the gap between home and school for the child. Vasudha, too, endowed the teacher not only with importance and authority but also with the responsibilities she must undertake:

> We do try our best to inculcate the best values . . . through our body language, through our own selves like the way we are talking, we try to project ourselves the best so that the children can take the best from

us . . . She [the teacher] has to be a role model. Whatever she's doing she has to keep in mind that the child shouldn't develop any kind of a fear for the teacher, she should be very patient towards the child, she should be able to understand the child.

She provided an example of how the teacher had to set the tone by her own behavior, which would then be modeled by the children: ". . . we've got a rule in the school bus that none of the children are supposed to eat in the bus, because our driver and conductor are very strict . . . so I make it a point that I don't eat in the bus myself because I've told my children not to . . ."

Malvika, on her part, highlighted the parent–teacher–student connection and explained, "when parents come for the open house they tell me that whatever way you [the teacher] behave in class, the children go home and they behave . . . They imitate each and every action that you do . . . There is a lot of respect. I mean, the way the kids respect the teacher the parents also respect the teacher. It's not only them [the parents] saying they respect us but they actually do whatever we say. And thank us for simple little things . . ."

Teacher–Child Interactions

With regard to interacting with children, disciplining children, and reinforcing the right attitudes and behaviors with them, almost all the participants I spoke with and whose classroom practices I observed demonstrated an approach that may be characterized as being honest and direct. In all of the observed interactions between teachers and their students, or even with their own children where applicable, there was an absence of the use of phrases that were sugarcoated or that instilled a false sense of praise. The "good job" phenomenon that prevails widely in classrooms as well as in the home environments of many children in the United States was noticeably absent. Charu, Malvika, and Vasudha seemed not to hesitate in pointing out when a child did something correctly or incorrectly. Rather than praising everything a child did, the feedback given was direct and to the point, the goal being to help each child recognize clearly their areas of strengths and weaknesses. The teachers were firm and strict yet kind and loving, and were comfortable speaking candidly to the children without fearing legal repercussions. In fact, this approach is quite contradictory to the approach of many parents and teachers in the United States who feel it is necessary to validate everything that a child does with praise so as not to damage the child's ego and self-esteem, which then might result in antisocial behavior. New (1999) notes that the fear of causing damage to the child's self-esteem has led to the practice in the United States of ensuring that children are provided more experiences in which they will succeed rather than fail, and it is in their best interest if adults don't ask too much of them. Recent research in the United States has indicated that high self-esteem does not necessarily prevent antisocial behavior, and, in fact, many people who pose a high threat to society are characterized by having high self-esteem (Slater, 2002; Sullivan, 2002). The widely held belief in the United States that self-esteem precedes healthy development may have led to overindulgence, and rather than focusing on the mastery of a skill many parents focus on reassuring their children that they are good in everything. The approach in India, on the other hand, is based on the belief that genuine self-esteem is developed in the child as the result of successful task completion, accomplishment, and skill mastery, and a false sense of praise does not facilitate this.

Malvika explained how both the home and the school had to help students work harder in areas that needed improvement: "we also seek the parents' help. We tell them that your child needs some extra effort, some extra work at home. Now they are in Class One [first grade] . . . and they're still doing their one page of cursive writing every day to improve upon their speed, their handwriting. So it's not only teachers who work but we also look for parents' help in this area, and truly speaking, parents are very co-operative . . ." The nursery teacher, Vasudha, too, echoed similar sentiments: "I think this [classroom work] is just an extension of what they're doing at home. We provide a guideline to the parents . . . and we also take guidelines from the parents . . . and with the help of that we study the background of the child . . . And we nurture the qualities accordingly."

For all three teachers, it was important to place the individual child in view of his/her duties and responsibilities. This aim did not seem to be rooted in a specific educational theory or philosophy but was connected more, in a subconscious way, to the overarching spiritual and social values of *dharma* in Indian philosophy. This was also representative of the wider body of early childhood teachers judging from the responses to the survey questionnaire that had been distributed. It may then be reasonable to say that teachers are influenced by their own sociocultural experiences to foster socio-emotional growth in the children in their classrooms, and use their knowledge of subject matter and expertise in content areas to help children achieve academic proficiency.

The Notion of the "Strict Teacher"

One issue that kept emerging in many of my conversations with the participants was their views regarding a "strict" teacher. At several points, Charu,

Malvika, and Vasudha all mentioned the fact that they had learnt a lot from a strict teacher or teacher educator, a strict parent or a strict mentor. It was interesting to note how each understood the notion of strictness as actually having facilitated their own learning process and helping them develop skills that were ultimately useful to them in their adult life and professional work. Charu spoke about the regimental schedule of her boarding school, which made her unhappy, but which, as she realized as an adult, helped her grow into a stronger and better person. Malvika shared that her mother was the strict parent, and her father was more like a friend. And yet it was from her mother that she learnt more and whom she turned to for guidance. Vasudha shared several examples of how she had learnt a great deal from her strict supervisor at work, and also from a strict college professor who had been her advisor during her teacher training. Most of these stories have been quoted in the book, but I want to share this one story that came from Vasudha:

> There was this lecturer, this professor . . . Every body was very scared of her. The first day when I joined, the first month, we were given our Tutorial groups. So unfortunately or fortunately—unfortunately at that time—she was my Tutor. I was paranoid. I didn't want to go to her. I was very scared. I didn't want to go to her class. She would just pick on anybody and—but that teacher turned out to be so good for us. I will never forget her in all my life. She was such a well-read person, and, I don't know, she encouraged me a lot . . . and she gave me so much encouragement, then I was never scared of her and I was very happy. She was a very good professor.

Charu reminisced about the years she was in a convent hostel when she was in grades six to ten: "Everything was on time. You had to study on time, eat on time. Eat what was served to you. Not like the way our kids behave. Things were really strict. We used to grudge it at that time. But now when I look back I feel that regimental kind of thing was really very important for me to grow up as a person." When I asked her to tell me more about her teachers in school, she added that they had been very dedicated: "We were a smaller number of children, and they knew us very well. And they were really dedicated. Completely. We had a lovely experience." Charu appeared to hold the same standards with regard to herself as a teacher as she explained, "you have to be strict for the benefit of the children plus you have to be accessible to them, open and friendly also so that they view you as somebody they can get across to . . ."

In my own understanding of each of their situations, it seems to me that their definition of strictness did not refer to a person who would inflict harm on the child, but referred to the fact that the adult in question held them to very high standards in their behavior and work, and expected no less from them. This, in their perceptions, ultimately helped them improve and refine their own performance. I got the sense that a strict teacher was not necessarily viewed as a bad teacher. In fact, it was important for a good teacher to be firm and be strict in setting high standards and expectations for her students. If the teacher was dedicated and committed to helping her students learn, expected the same commitment from her students in their work, encouraged them to work hard toward the set goal, and provided honest feedback and comments about their performance she would be viewed as a good teacher who had been instrumental in facilitating the development of many desired skills and good habits in her students. From the above conversations and from other interactions I had with people I met, it appeared that both good parenting as well as good teaching were defined not only by love for the children but also a certain degree of firmness and having strict expectations for children's behaviors and work habits.

Teacher as a Mother

Charu and Vasudha, both being mothers, viewed their students in the same way as they did their own children. In both their cases, aims of education for and attitudes and verbal interactions with students in their classrooms reflected the attitudes they had toward their own children. This also included the manner in which they reprimanded or admonished children. Vasudha correlated the connections between her own values and the values that she wanted her daughter and her students to learn:

> . . . if I'm honest, I want my children [students] to be honest. And the same is with my daughter also. The way I'll treat my daughter the same way I would like to treat my children as well. The same values I would like to be inculcated in both. I want to . . . give them [my students] the same education that I think is best for my own daughter. And my philosophy is that if I can love a child I can even be firm with the child. Because he is my child.

This approach seems to provide some degree of consistency in the adult–child interactions and in the expectations teachers and parents might have for children. Consequently, based on the belief that with the right to love a child comes also the responsibility to teach good values, the fear of reprimanding children is not a huge concern for most teachers in India. By and large, the expectations a teacher holds for her students is similar in many cases to parents' expectations for their children, and thus positions the teacher in a parental role. This in turn facilitates disciplinary issues because children are cognizant of the fact that their parents will, in most likelihood, support or respect the approach used by their teachers. Families demonstrate respect for the teachers and are desirous to facilitate and promote the academic and social objectives that the teacher works towards. Quite clearly, the teacher as a mother is a powerful notion that is taken seriously, and recognized not only by the teacher herself but also by her students and their families. One of the teachers who participated in the survey lucidly illustrated this point with her answer when she wrote, "a teacher has to become mother to hundreds and thousands of children that she adopts in her long career. It is a difficult job but of course a noble one." This approach of the teacher being like a mother is a contrast to the idea of professionalizing the teacher so that teaching can be viewed as a "real" job with a scientific and rational base, as opposed to the more homely and domestic job of mothering. But if the mothering is removed from the teaching, there is, I fear, a corresponding reduction in the teacher's sense of responsibility and accountability with regard to the overall development of the child's personality in accordance with acceptable social norms.

The next issue that I address is how matters of professional freedom, teacher's voice, and democratic classroom decisions play out in the teacher's work and practice when the teacher is perceived within a very traditional image.

The Teacher as a Professional

The teachers who were interviewed indicated that although they worked within the confines of a rigidly prescribed academic curriculum, they still had many opportunities to be themselves, to create within the curricular guidelines, to control approaches in implementing the curriculum, to make choices on which activities to include or which values to teach, to allow for various levels of noise in the classroom, to engage in collegial learning, and so forth.

Apart from being a class teacher, Charu also happened to be in charge of the four classes at her grade level. Each of the four teachers would take on that responsibility on a rotational basis. She said, "we had a very boring syllabus for *EVS* [environmental studies]. And I asked for permission if I could change it. So I changed it in such a way that every topic was related to the environment. Now we have plants one month, food and nutrition this month in January, weather, water, and so on . . . ." Charu indicated further that there were also several opportunities for her to make choices within the daily schedule in the classroom: "Usually we have about eight periods, that's about thirty to thirty-five minutes each. Initially we have a one hour long

Home Room period where the teacher is supposed to interact with the children. Of course, there are different days . . . like today they had Computers. And other days we're supposed to be with them and plan out different activities. So usually what I do is have an activity which continues or becomes a part of your curriculum."

Vasudha, the nursery class teacher, found even more space and time for shaping and planning the curriculum for her own classroom: ". . . I like to implement my own ideas. Whatever plays and songs we do I write on my own. If I've done one before, I will not repeat it. I have never repeated it . . . I write my own stories, my own characters. I go down to the child's level and everything is created by me. I love to do that . . . But I'm a very bad organizer. In eight years I've written so many plays, so many stories. Every year I write about ten to twelve. But I've never kept a single story!"

These images and roles actually fit in with the reflective model of teacher education that is proposed by Sparks-Langer and Colton (1991), who identify the cognitive, critical, and narrative approaches to reflection. In the case of the teachers in India, however, it was seen to happen informally during their practice rather than formally during classroom teaching or during their preparation in the teacher education colleges.

Teacher Training as a Colonial Construct

The concept of teacher training as a discipline was born in India only during colonization. The first institution to offer teacher training and to formalize the notion that teaching was a technical skill and teachers had to be trained was started in 1787 by Danish missionaries. By the year 1901, there were 155 teacher training colleges in the country. Thus, most of the teacher education institutions in India have colonial roots in European models. In all the years of the history of Indian education that can be traced as far back as 2000 B.C., there is extensive literature available that describes education in terms of philosophy, aims, structure of schools, curriculum, the teacher–student relationship, teachers' role and responsibilities, and so forth. Interestingly, there is hardly any mention of teacher training programs and the prevailing idea was the assumption that the best, the brightest, and most virtuous student would naturally take on the role of teaching.

According to the literature available, even before the start of colonial rule, school instruction in India was widespread in the form of an extensive system of popular education in many parts of the country, and teaching had been known in India to be a special form of social activity where teachers have been traditionally revered by all. Adam's survey of 1835 showed that teachers had the autonomy to choose what was worth teaching and how to teach it (Kumar, 1992; Goyal, 2003). Later, when a bureaucratic system of education was implemented by the British during colonization, it took away the freedom of teachers and denied them a voice in matters of the curriculum. The colonial government had the sole power to determine the syllabus, and design textbooks and teacher training courses. The teacher's role became confined to helping children learn the text that had been prescribed by the department of education's bureaucracy, and to maintaining logs and registers. These then were the skills that teacher education colleges, started by the colonial government, were expected to instill in would-be teachers.

Consequently, in India, teacher training developed as a Western construct, whereas the image of the teacher continued to reside largely within the Indian worldview. One of the components of practical knowledge is the image of how good teaching looks, and according to Elbaz (1981) teachers use their intuitions to fit into their image of good teaching. This culture-bound image still strongly influences not only the practice of teachers but also the hiring of teachers by school principals in India. More than the formal degree in early childhood teacher education, school principals look for teachers with high academic scores in content areas, pleasant demeanor, the ability to get on well with people, and the appropriate dispositions to impart values to their students—in the image of a traditional Indian teacher. These reasons were consistent with the responses of almost all of the school principals who participated in the study.

Criteria on Which Teachers Were Hired

The role and the image of the teacher in India are based on ideas and beliefs that are rooted in a cultural context rather than being explicitly informed by learning or developmental theorists. Since teaching about the role and image of the teacher is unattached to formal theory, it is absent in the teacher education curriculum. In the process of hiring teachers for private schools, the priority for the principals participating in the study appeared to be the search for teachers who fit this preferred socioculturally constructed image in addition to the mandated teacher education degree, certification, or diploma. In the survey questionnaire, the school principals were asked whether they thought teachers they had hired were appropriately qualified to teach early childhood classes. Some of the responses to the question were:

- "Not too often. Have to be given guidance through workshops, seminars and meetings in school."

- "They are good for N-2 classes because we select them on the basis of merit."
- "NTT (Nursery Teacher Training) better prepares for N-2 than B.Ed."
- "New teachers are hired only on the basis of past experience."
- "Primary teacher training is preferred but it is non-existent, most teachers go in for higher qualifications."
- "Very often no. Too highly qualified and can't come down to the children's level. So we hire them on the basis of good communication skills, creativity, positive attitude to learning."
- "No, most have a B.Ed which is not appropriate for N-2."
- "Yes, because we hire only qualified and experienced teachers."
- "Hire only experienced teachers."
- "Teachers are hired and chosen with great care."

These responses seemed to imply that teachers for classes Nursery through Two were rarely hired by school principals solely on the basis of having being trained in the B.Ed. program. Rather, principals hired teachers with great care based on their past teaching experiences, positive outgoing personalities, academic merit and creativity, and so forth. This might explain why the early childhood teachers hired by The School did not necessarily hold formal degrees in early childhood education. In one teacher's own words: "in India, I think you learn from your own experience and your background. A lot depends on that. How you were taught in school makes a lot of difference. As does your personal upbringing, your family background. I mean, that's what I observe when I see the other teachers in school also."

As indicated by the review of college documents, teacher education colleges in New Delhi offered two basic options: (1) the one-year-long Bachelor of Education (B.Ed.) program degree, which requires candidates to have a high school diploma plus a three-year bachelor's degree in a content area; and (2) the four-year-long Bachelor of Elementary Education (B.El.Ed.) program degree, which accepts candidates straight out of high school. Most teachers have a B.Ed. degree or a B.El.Ed. degree for the simple reason that teachers are mandated to have this degree in order to teach Kindergarten through Grade Twelve in New Delhi. The alternative for early childhood specialization is the nonuniversity-related diploma in Nursery Teacher Training (NTT) for women who want to teach in nursery schools. However, diploma holders are usually high school graduates who have not studied for a Bachelor's degree in a content area that, on the other hand, is a prerequisite for the B.Ed. degree. It seemed, thus, that school principals preferred to select teachers for the early childhood classrooms from candidates with university degrees even if their teacher education training is more appropriate for elementary grades rather than have teachers with only diplomas or nonuniversity-related early childhood training. The decision could be based on the belief that the candidates who are college graduates are intellectually more sophisticated, better prepared and qualified, having received a higher level of education, and possess higher-quality literacy skills with expertise in content areas, and thus are likely to be better teachers.

In addition, the fact that the curriculum for teacher education colleges is a psychology-driven one also raises the concern that although teacher graduates are well equipped with technical skills, the truth is that there is much more to teaching than instruction. Their teacher education program does not seem to prepare them in socialization skills and to teach children the cultural and values component of education, in which case they have to rely on their own instincts and upbringing. Consequently, many school principals in private schools in New Delhi were also inclined toward a stronger consideration of factors such as personalities, achievements, and upbringing of the candidates, during the hiring process of teachers, all of which contribute toward higher efficiency in teachers' socialization skills.

Socialization of teachers is the process whereby individuals acquire the values, attitudes, skills, and knowledge of the culture of teaching. The development of teachers' interpersonal skills and the above attributes is rarely considered in their professional training and thus their socialization skills are largely shaped by the teachers' backgrounds and local contexts. Teachers' cultural, racial, class backgrounds, and personal histories shape their cognitive frameworks or worldviews and thus influence their socialization; their personal background influences their selection of schools and their relationships with their students, and they learn about teaching from observing their own teachers (Achinstein, Ogawa, & Speiglman, 2004). Even in Europe and the United States, there has been recent but increasing interest in the development of interpersonal skills, social skills, social competence, emotional intelligence, and social knowledge of teachers (Stemler, Elliot, Grigorenko, & Sternberg, 2005). Some researchers even suggest that teacher education programs should expand their admission requirements beyond academic criteria to include personal attributes and dispositions of candidates and potential teachers (Chester & Beaudin, 1996; Marso & Pigge, 1997). Five attributes of effective urban teachers have been identified: sociocultural awareness, contextual interpersonal skills, self-understanding, risk-taking, perceived efficacy (Sachs, 2004). These attributes collectively (1) create a respectful trust between teacher and students because of the way the teacher views the students' life experiences as meaningful and valuable; (2) lead to enhanced communication between teachers and their students and families; (3) are conducive to strong collaboration between teachers and their colleagues and community; (4) can lead to a strong sense of self-inquiry between teachers' own fundamental beliefs and practice; (5) create individuals who are challenge-oriented and can act as change agents; and (6) facilitate the setting of high standards by teachers for themselves as well as for their students' persistence and assumptions of success (Gay, 1995; Guyton & Hidalgo, 1995; Ladson-Billings, 1995).

The above research seems to be pointing to a similar dilemma that is faced by the school principals in India as described earlier, and might also provide an additional perspective on why the school principals preferred placing a higher value on teachers' own personal characteristics rather than on their teacher educational degrees.

What Makes a Teacher "Good" and Successful

In concluding this chapter, I want to share some of the responses I received when I asked the participants during the interviews as well as in the survey what they felt were the most important qualities of a good teacher. The qualities most frequently mentioned were identified, and are presented below in order of decreasing frequency of the number of times they were mentioned:

- Being empathetic and understanding
- Being patient
- Being able to work with the individual levels/capabilities of each child
- Being loving, kind, or compassionate
- Being able to develop good character and love for learning in the child
- Having good communication skills

Empathy, understanding, and patience were by far the most frequently emphasized qualities for a good teacher. The image of a good teacher seemed to focus clearly on the affective qualities of the teacher rather than on her technical skills. Several participants indicated that the good teacher should have excellent communication skills and be able to work with children at their individual levels.

One interesting contrast was apparent between teachers' relationship with their students and families in the classrooms I observed in India and those I was familiar with in my work in the United States, specifically in terms of challenges to teacher authority. Due to the respect traditionally accorded to a teacher in India, a level of authority is automatically conferred upon teachers on the basis of their professional status. In a way, noncompliance by students or families is almost unthinkable in Indian schools and teachers are almost never criticized in the presence of children. This promotes a high degree of confidence on the part of the teacher who is able to address her students with calm confidence and free of hostility; when students are addressed confidently they are more likely to respond in the desirable way, and conversely, when the teacher addresses her students tentatively they are less likely to respond in the desired manner (Katz, 1999). In Euro-American contexts, however, the values of democracy, independence, and autonomy have been traditionally prized and this leads to a much higher degree of teachers being confronted and their classroom practices being challenged and negotiated (Alexander, 2000). Thus, teachers in American classrooms have to work harder and more actively establish their authority as compared to their Indian counterparts.

Judging from the information obtained from the conversations with all the educators, classroom observations of teachers, and survey responses from teachers and other educators, the general roles and responsibilities of an urban private school early childhood teacher in India may be summarized as:

- Balancing the academic and spiritual development of the students
- Teaching meta-academics and values as a parallel curriculum to the prescribed academic curriculum
- Providing direction and guidance to the students
- Setting clear boundaries
- Winning the love and respect of the students
- Treating students in the same way the teacher would treat her own children, whether in affection or reprimand, and be a motherly figure to them
- Possessing a high degree of expertise in academic content with in-depth knowledge of a content area
- Being a role model to reinforce values, behavior, and attitudes to children by first modeling it herself.

Ashwini, one of the teacher educators, shared with me her views of who an early childhood teacher ought to be and said she was "somebody who is enthusiastic, somebody who listens, somebody who talks, somebody who smiles, somebody who reaches out to children . . . someone who moves around in the class, children or students are able to openly and freely walk up to her, and somebody who is able to say that 'I don't know, let's look it up.' And somebody who can look at the mistakes and try and understand . . . mistakes children make tell us more about them and their learning skills than all the correct things they do. A teacher should be oriented to all that."

Chapter 6

Image of the Child: What Is Developmentally and Socially Appropriate for Children Growing Up in Indian Society?

A deeper understanding of children and childhood based upon an inquiry into the representation of childhood through mythology, cultural history, the history of spirituality, art, literature, and education results in an image of the child that symbolizes a culture's deeper assumptions regarding human nature, human forms of knowledge, and the meaning of the human life cycle (Kennedy, 2000). These understandings of how children and childhood are socioculturally constructed have implications on the day-to-day relationships between adults and children, and on educational theory and practice. Although it is not in the scope of this book to include an in-depth discussion of this type of research, it is, nevertheless, relevant to provide a general idea of the construction of childhood and how children are viewed in Indian society in order to contextualize the young child for whom the educational goals discussed earlier have been defined. My brief attempts to capture the image of the young child revealed an innocent being, immersed in an atmosphere of love, appreciation, and indulgence extended by the older members of the family and community.

The Hindi phrase for raising children is *palna-posna*, the meaning of which includes a combination of the concepts of cradling, nurturing, and protecting as opposed to the English terms of "bringing up" or "raising" a child. The young Indian child enjoys a longer period of infancy stretching from birth to almost five years with no other developmentally defined subdivisions. Traditionally, the idea of the child being born twice was commonly accepted in India, and there are still ceremonies in many Indian communities that mark the first birth, which is the actual biological birth of the infant, and the second "social" birth, which occurs any time between the ages of five and ten years and which symbolizes the child's separation from the adult–child unit and his birth into the larger community as an individual member of society. This extended period of childhood results in the general acceptance of a developmental continuum over which infancy and childhood are recognized for a longer duration of time. For instance, a child is not diagnosed or labeled as "developmentally delayed" if at the age of three years he/she exhibits the behavioral or developmental norms of a two-year-old as per the guidelines of Euro-American child development theories.

Although both boys and girls are loved and cherished by their families, there is a certain degree of rejection of girls that takes place in Indian society, especially within those families considered to be conservative and those who might be formally uneducated. This does not mean that there is overt and widespread hostility and violence toward all girls. Rather, it is the exaggerated importance that is given to boys that results in the creation of an indifference toward girls. This is one of the flaws in the social system in India, which is translated into a variety of behaviors and customs such as more intense and joyous celebrations on the birth, coming of age or marriage of a son; more inheritance rights to the son; more freedom to boys; more educational opportunities for boys; abuse of women by their in-laws amongst some families; and higher levels of infant mortality among females. Despite this, in many Indian homes the girl child is loved and protected by her family members and often thought of in the image of the goddess. Members of most communities recognize and honor the relationship between a brother and a sister in a very unique way and the popular festival of *Raksha Bandhan* has traditionally symbolized the lifelong love and protection that a brother promises to offer his sisters. My conversation with Malvika, the first grade teacher, revealed the love and protection that she as a girl received in her parents' home: "I've really enjoyed my childhood. See, one of the reasons I'm scared to get married is that I might lose all the freedom, that sort of love and care and everything . . . And I really want those feelings to continue in my life, and I don't want to lose them, and I want them to continue for a long time." She spoke also of how her role would be expected to change once she was married. Now, in her parents' home, she said "I'm so used to just taking and taking and taking, and there (in the home of her in-laws)—I'm already being prepared that you only have to give, give and give and nothing else, and don't expect anything from anybody. In fact others will expect everything from you . . . at times I feel I'm filled with love from within. I have received so much of love from each and every person I've known that I feel I have had enough of it and now I must give it out. I mean there are others who need it and I should actually pass it on . . ." Typically, young girls in India enjoy more affection and love, more protection, and more freedom in their parents' home and this changes considerably once they are married and begin to live with their parents-in-law.

Constructing the Image of the Child from Known Practices and Social Behaviors

It might be reasonable to say that the young child in India is considered to be a gift from God: energetic, mischievous, charming, lovable, intelligent, competent, playful, and certainly not a blank slate. Until the age of five years, this infant-child may be suckled, carried around, crooned to, snuggled, fed, clothed, and cleaned, all of which would be widely acceptable in Indian society. The child is filled with the belief that elders love and protect her/him, and the schoolgoing child is also told that her/his teacher is a very important and respectable adult in her/his life. Commonly heard endearments in Hindi-speaking regions, which parents and other adults use when referring to or addressing their children, are *mera raja* (my king) and *meri rani* (my queen), which indicates that a child in India holds the status of a very honored and deeply loved person. In fact, the words "*raja*" and "*rani*" are used so commonly that they often become a suffix following the names of individual children. On a personal note, I too have addressed my sons as "*raja*" since the time they were babies, and continue to do so even in their twenties.

A distinct adult–child continuity exists in Indian society whereby adults and children are a part of each other's physical space in daily life in a deeply integrated way, and enjoy each other's physical presence for a large part of the day. Having an infant sleep in the same room or even the same bed as the parents for the first couple of years is a common occurrence in India even today. This is in contrast to the popular American practice where families spend a large amount of time, energy, and money in furnishing and preparing a separate room for the unborn infant, and decorating the nursery where the newborn infant will be residing. In many Euro-American communities this is the first of many routines whereby a gradual physical distance between the child and the adults begins to occur. Another is the priority given to the reinforcement of self-help skills as even two- and three-year-olds are taught to be less physically dependent on their caregiver's help. Yet another step in the separation of the child from the adult occurs as children are encouraged at an early age to begin earning pocket money and this begins the process of children's financial independence from the parents. No doubt these practices teach children important skills and habits such as self-reliance, responsibility, and organization but these practices also create a degree of physical and emotional distance between adults and their young children. In Europe, the separation of childhood and adulthood can be traced back to events as early as A.D.1450, when the invention of "moveable" print and silent reading established a boundary between the inner and outer, or private and public, lives of people (Kennedy, 2000). A new definition of the publicly seen adult was created, and countless manuals on social etiquette were printed to reconstruct the definition of the modern, middle-class, well-mannered, and appropriately behaved adult. This new adulthood by definition excluded children and was therefore more distanced from childhood.

In India, adults and children are still physically a part of each other's worlds more intimately and for a longer time than is the norm in Euro-American communities. Of course there may be exceptions and the social norms of behaviors for children can be influenced by the sociocultural context of individual families and communities given the diversity of India. But, by and large, it is a fact that children are seen and heard throughout the house, running in and out of rooms freely even if there is a gathering of adults in session. Parents happily allow their children to sleep with them until the age of five years or even longer. Mothers go about doing their daily chores, holding onto their young children on their hip. I distinctly remember balancing my toddler son on my left hip, cradling him with my left arm, while I used my right hand to stir, fry, and pour while cooking; hung washed clothes to dry outside on the clothing line; tidied up a messy living room; or performed just about any household chore on a daily basis. Practices of childrearing in India may appear as though they encourage children to depend too heavily on the help of adults, and it may seem that children in India may not become as independent at as early an age as their American counterparts. Children allow their parents and grandparents to do things for them because of the pleasure it brings to the older generation. This is recognized and honored and dependency is not viewed as being negative. The larger picture of the human life cycle in India presents the adult–child relationship gradually evolving into one of interdependency. There is an underlying recognition that the lives of people are physically connected across generations, and that there exists a cyclical relationship between a child and an adult. Even after having grown up, children of most Indian families maintain their intimate connections to their parents, looking after them and providing for them in their old age. This concept of interdependency, in fact, leads to closer family bonds and stronger relationships between members within a family or a community. More importantly, although interdependency is promoted within the family the child is definitely encouraged to develop skills of independence and self-reliance outside the family. Flexibility is an interesting fact regarding childrearing in India and children are taught at a young age that different situations demand different skills and behaviors, and that the individual is called upon to play multiple roles and responsibilities over the course of his/her life.

There are several practical indications of how childhood is perceived as lasting for a longer period of time in India. One example can be seen in the rating of movies in India. Typically, movies made in India are not categorized in their ratings to cater to children in several specific age groups as they are in the United States, where the age groups are determined by an understanding of child development through Euro-American standards. The movies in India have traditionally been labeled either "Adult" or "Universal," with the former category restricted to individuals above the age of eighteen years, and the latter for everyone including young children. The decision as to which movies children are allowed to view is largely left to individual families and parents. Another indicator of childhood being viewed as a continuous period and not subdivided is seen in the manner of how children of all ages play together. Older and younger siblings and their friends play common games, caring for and looking out for each other; they do not have play dates arranged for within their specific age groups. Although structured and supervised age-appropriate groupings facilitate social interactions within those groups, spontaneous play occurring naturally within mixed age groups enable children to develop a broader range of social and emotional skills as they must now learn how to socialize with people of different ages, learn behaviors modeled by their older siblings and friends, and at the same time learn to model behaviors for those younger than them in the group. Children playing with each other and engaging in extensive verbal interactions is a sight seen more commonly and widely than children playing with objects. Knowledge as the process of social construction is very much evident among children growing up in India.

The contextual understanding of childhood in India starts with the image of the mother. Motherhood in Hindu society is revered as a moral and religious ideal. The very word for strength, *shakti*, is the name of the female principle *Shakti*, the counterpart of the god *Shiva*, who represents destructive energy. (The creative energy is represented by the god *Brahma*, and the god *Vishnu* represents the energy that preserves the balance between creation and destruction, life and death, or good and evil on this earth. All three together represent *Brahmaan*, or the core reality as the eternal source of all energies.) The words "strength" and "mother" are often used in juxtaposition implying that a powerful force is associated with a mother's love energy. The Indian baby is born within these conditions and is nurtured and protected by the strength of not only a biological mother, but all the other mothers in the extended family as well. The Indian woman in her role as mother focuses all of her energies on her child, from the fury of her wrath directed toward any who would harm her child, to the tenderness of her love directed toward the child itself. In the absence of the mother, the child's grandmothers, aunts, older sisters, and even maidservants are allowed to readily slip into the mother's role. The child is rarely ever left alone and is carried around by the mother or another caregiver as they go about their work. This child, then, thrives in a richly emotional and caring environment regardless of the nature of the socioeconomic, ethnic, or religious environment within which the family might exist.

Constructing the Image of the Child from Texts

Indian psychology, based on philosophy and mythology, understands infancy to encompass a wider age range than does Western developmental psychology. Numerous examples are provided in literary, philosophical, and devotional texts of how childhood in India has been historically viewed. These include, among many others, images from various Indian texts such as ancient law books (*The Laws of Manu*); the caring and upbringing of infants and children as prescribed in traditional medicinal texts (*Ayurveda, Charaka Samhita, Sushruta Samhita*); from references, stories, and narratives on children and childhood in the epics (*Ramayana* and *Mahabharata*); from folktales and historical narratives; and from a plethora of ancient, medieval, as well as modern Indian literature. Included in all these texts is detailed information on child development, the relationship between childrearing and personality development, intense parental longing for children, and children's upbringing that has often been marked by affectionate indulgence and divine protection. The adult characters that are mean or unkind to a child are most often the demons or the villains in the story and many stories end with the child escaping to a happier or more powerful existence. When the evil king Kans tries to kill his newborn niece by flinging her on the floor of the prison, she slips out of his hands and into the sky and turns into a bolt of lightning; when the young Prahlad is accused of not recognizing the evil king Hirnakush as being god, the king orders him to be killed by having him tied to a red-hot pillar. Prahlad's god sends him a sign of reassurance by showing him a tiny ant climbing on the pillar just before he is tied to it indicating that the pillar is not really hot, and the king is shocked that this young boy does not die. When the young Krishna is tricked into suckling from his evil aunt who has smeared her breast with a poison in the hope of killing him, he draws upon divine powers to suckle her until she is drained of all blood and herself dies in the process. In the approximately 1,000-years-old *Bhakti* literature, which is the most power-ful surviving literary tradition in Northern India, the child is described as an exalted being who grows up surrounded by a circle of admiring adults, is completely absorbed in an interplay with his mother, and is loved for the childlike qualities of freedom, spontaneity, simplicity, and delight in the self, which are considered as divine attributes (Kakar, 1981). Ingrained in the adult worldview is the commonly used phrase, "a child is one of the forms of God," and the refrain of a popular cinema song can be translated as "children are the flowers that are closest and dearest to God." There are endless such stories and references that reinforce the notion of a child as someone who is blessed, protected, and loved within a prolonged child–adult relationship. This approach is markedly different from the Euro-American Puritan influence on childrearing, which, in seeing the child as being born evil and feeling the need to impose strict discipline upon young children, resulted in the writing of several books and essays on the proper way to train young children to be quiet, obedient, and submissive. Many subsequent educational philosophies in Europe and the United States were defined by this belief.

According to the 5,000-years-old tradition of Ancient Indian medicine, *Ayurveda*, life begins at conception and not birth. There is a story in the *Mahabharata* of the yet unborn Abhimanyu, son of the great warrior Arjuna, who overhears his father disclose a secret war strategy to his mother while still in her womb and later is able to defeat an enemy formation during the mighty battle of Kurukshetra. The infancy stage according to Indian thought typically ranges from birth to at least four or five years (Kakar, 1981; Vyas, R.N., 1981). As a result, there is an extended period of intense attachment between the child and mother that is most clearly man-ifested in their physical closeness. The principles of *Ayurveda* define five childhood periods: *garbha* (the fetal period); *ksheerda* (birth to nine months) when the infant lives entirely on milk; *ksheerannada* (nine months to two years) during which weaning takes place and the child moves from milk to solids); *bala* (two to seven years); and *kumara* (seven to sixteen years). These periods are defined by stages that mark the gradual physical separation between the infant and the mother rather than by milestones in the child's individual socio-emotional development. Kakar (1981) describes some of the social rites of passage at specific ages that mark the child's journey through and out of childhood. A month after birth, the naming ceremony, *naamkaran*, marks the formal introduction of the infant and new mother to the larger family. At the age of three or four months the baby is taken outdoors to be introduced to the world and the cosmos as per the ancient texts during the ritual of *nishkramana*. Between six and nine months is the ceremony of *annaprasana* that marks the onset of weaning and the beginning of a psychological separation from the mother by feeding the infant solid food for the first time. At the age of three, the child's baby hair is shaved off and offered to the goddess in a ceremony called *mundan*, which is yet another step marking the gradual separation of the child from his mother. Between the child's fifth and seventh years is the ceremony called *vidyarambha*, which marks the child's readiness to start formal learning of reading and writing. The period of childhood culminates in a special ceremony or *upayana*, which is performed around the child's twelfth birthday and symbolizes the child's social birth into the larger community outside the extended family. Variations and combinations of these rituals and rites exist throughout Indian society, and can be seen to varying extents across social castes and classes in rural and urban communities.

Modern Indian educators and psychologists support the idea of a continuity between adults and children as being a part of each other's worlds for an extended period of time (Kakar, 1981; Kumar, 1993). As already described, this is seen widely when children continue to sleep with their mothers, continue to be suckled, and are carried around as their mothers tend to their household chores. The constant contact as the child is cuddled and crooned to leads to a prolonged experience heavily charged with emotion. The most common interpretation of a child crying when it is not in hunger or pain is that it is in distress due to being physically separated from the mother. During this distress, which is the result of a longing for her presence, not even the baby's father can be of comfort to the child. Since the child is still considered to be an infant for a longer period of time, there is no subsequent hurry or expectation to have the child learn to be more independent in terms of toileting, feeding, cleaning, self-care, or other self-help skills. It is believed that these milestones will be reached in due time, and sometimes sooner than expected because none of the emotional conflict in trying to make it happen occurs between the child and the caregiver. It is only around the age of four years that the real separation between the child and the mother begins, and the child starts to move away from the nurturing unit that he/she was a part of. Thus young children are nurtured, and not reared, raised, or molded to a set of adult expectations.

The quest for uniting with the ultimate spirit has always been the goal of Indian philosophy, religion, and life; and the entire scheme of human development has been organized according to this goal. Dave (1991) describes the four main values of life encompassed within the Hindu concept of *ashramadharma* as: (1) *artha* (money or sustenance); (2) *kama* (love and reproduction); (3) *dharma* (holding on to moral codes of conduct); and (4) *moksha* (final union with the ultimate spirit after release from the cycle of birth and death). The span of human life has also been systematically demarcated into developmental stages leading to the ultimate goal. These stages are *brahmacharya, garhasthya, vanaprastha,* and *sanyasa.* Kakar (1981) outlines a comparison between Erik Erikson's psychosocial stages of growth and the more philosophical Vedic concept of *ashramadharma,* or the stages of human development as understood within the Hindu worldview. Erikson's theory of Psychosocial Development plays an important role in the understanding of child development in Euro-American cultures. But unlike Erikson's theory, which recognizes the development of a healthy personality if the tasks at various stages are resolved successfully, the theory of *ashramadharma* does not focus on implications of mental health if the tasks remain unfulfilled. Instead, it emphasizes the importance of the progression from stage to stage in the purpose of life and in the ultimate realization of *moksha.* Figure 6.1 provides a summary of the comparison between Erikson's stages and those according to the Hindu scheme (Kakar, 1981).

Erikson's Scheme		Hindu Scheme	
Stage	Specific Task and "Virtue"	Stage	Specific Task and "Virtue"
1. Infancy	Trust vs. Mistrust: Hope	Individual's "prehistory" not explicitly considered	Preparation of the capacity to comprehend *dharma*
2. Early Childhood	Autonomy vs. Shame, Doubt: Willpower		
3. Play Age	Initiative vs. Guilt: Purpose		
4. School Age	Industry vs. Inferiority: Competence	Apprenticeship (*brahmacharya*)	Knowledge of *dharma*: Competence and Fidelity
5. Adolescence	Identity vs. Identity Diffusion: Fidelity		
6. Young Adulthood	Intimacy vs. Isolation: Love	Householder (*garhasthya*)	Practice of *dharma*: Love and Care
7. Adulthood	Generativity vs. Stagnation: Care	Withdrawal (*vanaprastha*)	Teaching of *dharma*: Extended Care
8. Old Age	Integrity vs. Despair: Wisdom	Renunciation (*sannyasa*)	Realization of *dharma*: Wisdom

Figure 6.1   A comparison of the stages of development in the schemes of Erikson and those according to the Hindu scheme. Reproduced with permission from Oxford University Press India, New Delhi.

Stages in the human life span as understood within the Hindu worldview may be broadly summarized in the following manner:

- The concept of *dharma* and duty is the core around which the Hindu life is lived. Stages in human development are recognized and defined by the understanding and practicing of *dharma*. Hindu philosophy reiterates the fact that each individual is born in his own *dharma* and must live by his own *dharma*. There is no one single set of rules that can be applicable to every human being. Each must understand one's own life-world and live according to the priorities and responsibilities of one's own *dharma*. This not only contextualizes a Hindu's life but also highlights the complex multiple realities lived individually by each child and adult.
- Although childhood within itself has many stages as described in *Ayurveda*, these stages mark the gradually shifting relationship between the child, his/her mother, and the larger society around them rather than the child's individual physical and socio-emotional development.
- The prime duty of a school-aged child and youth (*brahmacharya*) is to attain knowledge of the external world as well as getting to know one's own self and *dharma*.
- After the study period is over, the youth enters manhood and the domestic stage of life (*garhasthya*) during which the emphasis is on successfully establishing a home and raising a family.
- Middle age (*vanaprastha*) brings with it a completion of household responsibilities, and the start of withdrawing from the material world in search of inner truths that will point the way to *moksha*.
- Old age is equated with reflection and wisdom and the objective of completely letting go of the distractions inherent in the materialistic world *(sanyasa)*. This understanding automatically brings to elders a high degree of respect and veneration across Indian society.

The system of beliefs and values associated with the fundamental concepts of *dharma* and *karma* forms a reality for Hindus that is culturally specific and constitutes the identity of the Indian community. This is not a system of abstractions that must be comprehended during the adult years, or studied as a course on philosophy in a school or college classroom. It is a part of the Indian psyche, absorbed and imbibed by children from the very beginning of their infant years from their relationship with their mother and other adult caregivers as an understanding of the reality of the world in which they will spend their lives.

"Appropriate" Pedagogy and Practice

The image of the child constructed above naturally appears very different than that of the image of the child in the United States. The guidelines on Developmentally Appropriate Practices (DAP), which recognize a universal set of predetermined stages of human development, and define acceptable educational practices as referring to specific teaching styles, learning styles, adult–child interactions, classroom set-up, classroom materials, and so forth was based on developmental milestones and skills that were identified as being important for children in the United States. The danger of using the term "appropriate" is that any teaching or learning style that does not fall within this acceptable domain is excluded and marginalized as being "inappropriate." This then completely ignores the validity of the image of the child, notions of childhood, childrearing, and educational practices in cultures and communities that differ from the white, middle-class American way of life. DAP thus promotes a child-centered, play-based, activity-oriented classroom approach, which assumes that appropriate learning for all children can happen only through the methods it promotes, namely discovery and object manipulation. But it is a known fact that children exhibit different learning styles and construct knowledge in several different ways for themselves. These learning styles may include not only learning from experimenting and manipulating, but also styles that involve assimilating information by hearing someone speak, learning from reading print, practicing by writing and rewriting, comprehending by reading aloud, memorization of basic concepts in order to problem solve more advanced concepts, and so forth. Even the notion of play according to DAP is characterized by mostly physical forms of play and physical manipulation of objects and disregards any forms of verbal interactions and exchanges as play. It is important to keep in mind that there may be a place for multiple teaching and learning strategies in the same classroom.

Taking the DAP-based dominant early childhood discourse at face value, there are many educators in India who buy into the desirability of the American progressive, developmentally appropriate, child-centered, play-based pedagogical approach, and I would like to caution those who would strive to implement a purely Western child development approach into the understanding of early childhood education in schools in India. Canella (1997) makes note of the fact that the tenets of child-centered pedagogy and play have been created within a specific culture and a particular set of values and biases and are not the norm in *all* early childhood schools even in the United States. In fact, current early childhood classroom practices in New York City public kindergartens are becoming increasingly defined by academic mandates and high-stakes assessment and it is almost unusual to find child-centered, play-based approaches in New York City kindergarten classrooms. Applying this idea of child-centered practices to all people in all situations denies the existence of diverse cultures and knowledge reper- toires, with multiple worldviews, values, and learning styles. I would also like to raise the issue that in many early childhood classrooms, teachers impose their definitions of play without realizing that their students have their own understanding of which activities they consider as play or work. Even after making the distinction between work and play activities, King (1979) notes that the kindergartners who participated in her research did not themselves necessarily view the work activities as being bad even though the adults did. A fellow teacher educator who teaches a research course presented her student's study on the uses of manipulatives in an elementary classroom in a New York City public school, and shared findings that not all children in the classroom viewed the use of manipulatives as facilitating their learning (Making changes, making knowledge . . . , 2005). Thus, child-centered pedagogy is a reflection of Western middle-class assumptions on the nature of learning, and middle-class values where success is depend- ent on the availability of materials and financial resources. The imposition of such pedagogy on all children would be equivalent to the colonization by Western progressive values of classrooms all over the world (Canella, 1997). Practitioners in India should be mindful of the fact that context and culture-specific norms that guide children's growth and behavior in India are different from children's behaviors in the United States, which are guided by the American progressive ideals of independent decisionmaking, individu- alized growth and development, learning to choose from several given options, engaging with an abundance of materials, voicing and expressing their feelings, creating their own stories, and so forth. It is important for educators in India to acknowledge the image of children and childhood as constructed within the Indian culture and worldview, and the developmen- tal and social behaviors that are more acceptable for children growing up in Indian society, thus carefully selecting what Indian schools can focus on in their own curricula. Some of these developmental and social norms are dis- cussed in the following section.

The Social Self

The concept of the self on the "self–other" continuum is located differently in the Indian context as compared to the understanding of the independ- ent, egoistic self in Euro-American psychology. In the Indian worldview, the self is held to be of a more social nature, and the "me" and "mine" is subsequently held to be secondary to "we" and "ours." Personal work is commonly set aside when visitors drop by, and there is a definite shift in the focus of the sociocentric self from its own ego to the caring and considera- tion of the "other" in the family and the community. In a classroom scenario, if a student is approached by another child, he would be expected to share his/her toy since not sharing would imply that the former is being selfish. In contrast, as has been noted by researchers, classroom practice in the West, which is based on child development theory, embraces a pedagogy that maintains the importance of the egocentric self, and encourages the individualization of this self (Jipson, 1991; Kamii, 1984). In a typical pro- gressive classroom, the approaching child would be asked to wait until the first student has completed his/her play with the toy. In accordance with progressive educational and child development ideals, good teaching would be defined as meeting the needs of the individual child and this further highlights the egocentric nature and overlooks the sociocentric nature of the child.

The Teacher–Child Relationship

This has been described in detail in previous chapters, and the notion of the adult–child continuity has been discussed earlier in this chapter. Generally, in an Indian classroom, the teacher is viewed as more experienced and more mature than the young child. Hence "telling" the child what to do is part of the teacher's duty, just as it is the responsibility of the adults in the child's family. Guiding the child during formal teaching, and even modeling for the child is actually recommended. This approach is more in keeping with the present strategies, which are allied with the Vygotskian-informed theo- ries of scaffolding (Wood, Bruner, & Ross, 1976) and guided participation (Rogoff, 1990), to facilitate children's learning and children's development in a sociohistorical and sociocultural context. Research indicates that this deliberate form of guided participation at a young age actually increases the child's ability for independent work at an older age (Rogoff, Mosier, Mistry, & Goncu, 1993).

   This active guidance by adults and older peers has been historically less acceptable in the progressive early childhood classrooms in the United States, in which practices have been strongly influenced by the Piagetian and Eriksonian approaches underlying the recommendations of the DAP (Bredekamp, 1987). Although this set of guidelines was recently revised (Bredekamp & Copple, 1997) early childhood classroom practice nation- wide is still driven by the tremendous momentum set in place by the first edition because it was endorsed by a powerful early childhood organization, the National Association for the Education of Young Children, and was used as a basis to grant accreditation to early childhood centers and schools across the United States (Williams, 1994). In many such progressive early childhood classrooms, modeling for children has been traditionally discouraged; following a more Piagetian approach, the child is left to initiate his/her own learning and problem solve situations for himself or herself on a path toward increasing autonomy (Bredekamp, 1987; Walkerdine, 1984). The intervention of an adult, verbal or physical, has been widely considered to be detrimental to the child's development of initiative and creativity.

The fact that in India an adult can "tell" the child what is right or wrong in terms of what is acceptable or not, does not necessarily imply that the children have no opportunity to make individual choices, or that adults want to exert total control over them. The participants in my study frequently reiterated that it was important for children to express their likes and dislikes and make some of their own decisions (Malvika); to develop into autonomous persons and learn to think for themselves (Nagma and Ashwini); that their minds should grow and not be stumped by too much authority (Shakti); or that children should feel that coming to school is fun (Veera). There were several other examples illustrating that most of the participants viewed a good balance between teachers' guidance and children's independence as being the appropriate approach.

Leadership Skills

In India, the multiple kinship roles of individuals require them to exhibit the appropriate social norms and behaviors that are specifically defined for each role. The particular nature of interactions between individuals thus keeps shifting depending on which persons are participating in the interactions and how they are positioned in the family hierarchy. It is understandable, then, that leadership skills do not assume high importance and an individual is not assessed on the basis of strong leadership skills. It is equally important for individuals to be able to follow instructions and guidance. What becomes important to learn is how to balance these contrasting skills and be flexible in being a leader and a follower in a given context, and maintaining harmony within a larger group.

Verbal Skills

Similarly, in India, verbal skills are not a measure of accomplishment in growth and development, but task completion and following up on professed goals are. Individuals are assessed on actual performance and the results of a task rather than on self-articulation of their own skills and abilities. The need to "sell" oneself is minimized thus reducing the need for verbal presentations. How an individual is evaluated by others has been considered to be more important than how the individual perceives himself/herself. However, this does not imply that verbal skills have no importance in children's lives. Indians by nature are very chatty people with a rich oral tradition, and the play of children is by far more verbal than the play of children in the United States, where play is characterized more by the physical manipulation of and engagement with objects. Play amongst children in India is strongly marked by interactive skills such as teasing, beseeching, counting, rhyming, touching, running, and so forth.

Academic and Social Learning

Since Indian society is characterized by continuity between childhood and adulthood, and by a very vast and detailed kinship system within which numerous kinship titles distinguish the numerous ways in which any individual is related to others in the community, there is a high degree of human interaction and the power of the extended family concept cannot be underscored enough in Indian society. Maintaining the relationships between kith and kin is the one most important duty and responsibility for an individual. Even as more families become nuclear in modern times, the unseen connections are as strong as ever with family members still visiting and supporting each other, and protecting the weaker and more vulnerable members (Dave, 1991; Kakar, 1981). To a large extent, much of a child's social, emotional, and moral development takes place as he or she lives within this huge network of families, relatives, and friends in an environment that is bustling with numerous caring, nurturing, loving, and reprimanding adults modeling social norms and behaviors for the child. It is important to know that Indian adults do not create learning situations to teach their children, but rather young children always being in the presence of adults, learn from watching them go about their daily activities (Yunus, 2005). Thus, social norms and customs are learned behaviors in India, rather than actively taught behaviors. A value such as respect for elders is learned by Indian children as they observe their parents caring for and being respectful to their elders. Charu, the second grade teacher, mentioned honesty as a value she prioritized. When I asked her how she was taught honesty, she seemed surprised and said that her parents never explicitly taught her about honesty. She learned it from their behaviors and the way they conducted themselves. With such a high degree of socio-emotional development occurring in the home environment, the school is automatically positioned as an institution where the child must learn formal academic skills. Therefore, the socio-emotional development of children falls more into the domain of home and family, whereas academic learning falls into the domain of schools and classrooms. It becomes possible for teachers to focus at length on the prescribed academic syllabus, which, in turn, facilitates the implementation of high academic standards and high-stakes assessment. This is, perhaps, one of the reasons why and how the examination driven system has continued to exist in India.

Attention Span and Focus

Children growing up in India are expected to develop longer attention spans and greater attention to detail at a younger age than is recommended by child development guidelines in the United States. Developmentally appropriate progressive guidelines discourage classroom practices in which young children are expected to sit for longer periods of time, and physical movement in the class is greatly encouraged (Bredekamp & Copple, 1997). Progressive early childhood classrooms in the West are therefore designed to allow such physical movement by the creation of various activity centers in the room and by opening up more space through the elimination of desks and chairs for the students. The students are expected to walk around the room from center to center, and even stand while working at some centers. This level of importance given to physical movement is understandable in a society that values sports and physical activities. On the other hand, children in Indian schools and society are expected to learn how to be closely attentive and focused, and sit for longer periods of time. Scarcity of resources, space, and materials teach children early to wait for a turn and share their possessions. Physical movement within the classroom becomes impossible when there are 40 students and as many desks and chairs. Although physical energy is not evident in classrooms in India, there is a higher level of verbal and mental energy. I have discussed the manifestation of energy in different forms in more detail in chapter 8, which explores the issue of class size. Cultural tools to encourage focus, stillness, and attention to detail are built into traditional practices. One example is that families recognize and celebrate milestones in life and other accomplishments by performing a *puja*, or prayer ceremony. Typically these ceremonies are performed on the ground, with members of the family and the priest seated on the floor. Young children participate along with the adults, and everyone is expected to remain seated for long periods of time while listening attentively to the priest's prayers. Of course, very young children are free to get up and walk around, but many of them gradually do begin to model the behaviors of their parents and grandparents. A few months ago, on my last visit to India, my family held one such prayer ceremony, which was attended by children and adults across four generations of my family. The children included my six-year-old nephew, born and brought up in Seattle and who by nature is an active, energetic, and highly curious little boy. I was surprised to see him stay seated quietly on the floor throughout the duration of the 90-minute-long ceremony. The silent expectation was there, in addition to the fact that he so wanted to do what his older cousins were doing! At the end of the ceremony, all of us, and he most of all, expressed immense pride in this accomplishment.

Fine Motor Skills

Closely related to the point above are the differences I observed with regard to the development of fine motor skills. According to developmentally appropriate practices in the United States, the use of thin pencils, crayons, and paintbrushes are discouraged for children under the age of four years. In India, fine motor skills are developed in children at a younger age because they are encouraged to focus on and become attentive to detail through the use of smaller and finer writing tools and paintbrushes. An example of another cultural tool that aids the development of fine motor skills in Indian children is the technique of eating food using fingers. Indian children learn to eat with their fingers at an early age and the process involves a highly skillful manipulation of food with the fingers of one hand. In contrast, children growing up in countries where the use of utensils is preferred, either use finger foods that are simply held and bitten off, or use a spoon and fork, which requires wrist and elbow movements rather than the movements of the smaller muscles of the finger digits. Teachers in early childhood classrooms in the United States accordingly expect three- and four-year-olds to use thick crayons, thick pencils, and large brushes for writing and easel painting, which require the use of wrist and elbow movements primarily. In Indian classrooms, in the absence of resources such as thick pencils and easel painting materials, young children learn how to use regular pencils and finer paint brushes at an earlier age. Interschool art competitions are commonplace and very popular all over India. Even kindergartners are encouraged to participate in these city-wide contests in which hundreds of students gather at one venue, armed with their drawing papers and crayons, and sitting on the ground they draw pictures with the most amazing details. The drawings are then evaluated and judged, and prizes are awarded to the best three or four. Many of the sheets of artwork make their way into calendars and various displays in schools. As a young mother, my pride knew no bounds each time my younger son was enrolled by the art teacher in his school in many such competitions, where he won several prizes.

Independence vs. Interdependence

Emotional closeness, display of feelings, and physical interdependence are valued in India. Children are considered to be a gift from the gods and are thus nurtured and protected by all the women in the family (Kakar, 1981; Vyas, R.N., 1981). Therefore it is completely acceptable for a four-year-old to be carried, cleaned, or clothed by a caregiver in India, whether at school or at home. However, child development research in the United States and Europe, which defined the widely implemented, original, developmentally appropriate practices considered it inappropriate for teachers and parents to "baby" the child, who is expected to be more independent and self-reliant at the age of three or four (Bredekamp & Copple, 1997), or is otherwise at risk of being labeled as developmentally delayed. Interdependence between different generations is a reality in the lives of Indian families, and there is no rush for children to become as independent as possible at as early an age as possible. I have discussed the concept of interdependence earlier in the chapter, where I highlighted that interdependence actually led to stronger family relationships across generations.

In conclusion, and as a word of caution, it is important to consider that developmental and social norms be taken as patterns and trends within cultures and communities, but not be rigidly generalized as rules. When studying developmental characteristics across cultures, it is important to remember that these can vary vastly between even closely related communities. One mistake that has been made often by educators and researchers is that Asian cultures seem to be grouped together. Asia is a large continent that is divided into specific regions such as South Asia, East Asia, Southeast Asia, and so forth, and includes countries as diverse as China, Japan, Indonesia, Malaysia, Korea, Vietnam, India, Pakistan, Afghanistan, among many others. Not only are differences in attitudes and behaviors evident between Indian and other Asian cultures, but differences will also be apparent between different communities within Indian society. However, without doubt, the cultural differences in how children are viewed and expected to develop in the United States and the corresponding pedagogy that defines teachers' practice in American schools is quite different from that in India. So rather than drawing upon child development theory, which is based on research that has been conducted in culturally different contexts, my primary recommendation is to encourage educators in India to assemble a body of formal research and literature on child development and pedagogy that would be more culturally relevant for schooling in India and which could eventually be included in the formal curriculum of teacher education programs that prepare teachers for Indian schools.

Chapter 7

Learning to Teach: A Sociocultural–Historical Constructivist Theory of Teaching

Most teacher preparation programs around the world have been generally based on the assumption that student teachers after having been informed about research-based theoretical structures will graduate from their institutions, enter classrooms for young children, and will simply apply these theoretical structures to practical situations. This approach is less supported now, and there is more research to support the fact that teacher effectiveness depends not only on cognitive outcomes but also on students' moral and social well-being, as well as on the establishment of positive relationships with colleagues and parents (Campbell, Kyriakides, Mujis, & Robinson, 2003). But there is still a marked absence of a theory of teacher cognition and behavior in teacher education programs that formally includes such concepts as teachers' own personal theories and practical knowledge (Korthagen and Kessels, 1999). Further, although there has been increasing recognition for the importance of good interpersonal skills and social knowledge as a mark of effective teaching, the systematic development of such socialization and interpersonal skills and strategies is relatively absent from teacher education curriculum, and how teachers react to different social situations in the context of their teaching is usually dependent on their own personalities, background, and experiences (Mills, 2003; Stronge, 2002).

I briefly review some previous research that has established that teachers construct knowledge in various ways through their sociocultural experiences and bring this body of knowledge, comprising practical knowledge and their implicit beliefs, into the college classrooms as well as into the school classrooms in which they teach.

How Teachers Learn

Jalongo and Isenberg (1995) have listed the different ways in which teachers can be aware of something: formally through professional readings and studies; directly through their own classroom experiences; vicariously through observing and listening to the experiences of other teachers; and intuitively through their own value systems. Teachers' values represent the teacher's life goals, and constitute a theory of the world of teaching. One of the findings in Hollingsworth's (1989) longitudinal study of 14 preservice teachers entering a teacher education program in the United States was that prior beliefs greatly influenced learning to teach and that it was critical to use the understanding of preservice teachers' prior beliefs to inform university course design. Based on two case studies on the teacher as a dilemma manager, Lampert (1985) defined a classroom dilemma as a problem forcing a choice between equally undesirable alternatives, and concluded that who the teacher is determines to a large extent how she approaches classroom dilemmas, where the dilemma may be equated to making a moral and ethical choice with the decision to act coming from within the teacher and her sociocultural belief system.

It is widely accepted that teachers have considerable experience prior to entering the teaching profession and much research has been conducted on this topic. Spodek's (1988) review on teachers' perceptions of education strongly indicated that teachers' implicit theories are shaped by not only child development theories, but also by their own values and knowledge of traditions. Based on the findings of a case study, Clandinin and Connelly (1986) suggested that teachers acquire practical knowledge through an interaction between their own personal narratives and particular situations. However, it is seldom formally or professionally acknowledged that teachers have a repertoire of knowledge and skills based upon that experience. According to Clandinin (1986), the problem is that teacher educators fail to recognize knowledge that is more practically oriented. If belief systems that form the foundation of such intuition and practical knowledge were to be made a part of the theory that teachers are exposed to, then the role of practical knowledge would be a more accepted one.

Clark and Peterson (1986) drew attention to the psychological context of partially articulated theories, beliefs, and values in which a teacher works, and which defines her attributions for students' performances, teachers' roles, expectations, classroom behavior, importance of students' interests, how and what they learn, and so forth. Research by Olson (1981) also indicated tensions between teachers' implicit beliefs and curriculum developer's beliefs, with the teachers modifying a new curriculum and teaching it in a way that was more compatible with their own implicit beliefs of good teaching. Similarly, Schoonmaker and Ryan (1996) described how top–down, formal theories of education and learning presented in teacher education programs are diluted and modified by teachers' personal constructs, if an attempt is made to implement them in classroom practice. Elbaz's (1981) research on the practical knowledge of teachers suggested that this practical knowledge is not acquired through teacher preparation courses but evolves over time through their field experiences. One of the components of practical knowledge is the image of how good teaching looks, and according to Elbaz teachers used their intuitions to fit into their image of good teaching. Richardson (1994) made note of the fact that teachers' thinking was defined by context and experience, and was not like the linear and prescriptive planning that is recommended by formal research in teacher education programs.

Several attributes of teachers as indicators of their potential success have been identified by theorists in urban education, and these characteristics may include beliefs or perceptions about self and others, personal values, and morals that are held as a standard to guide one's thinking and behavior (Sachs, 2004). Based on a large body of research on teacher education, many recommendations have been made to connect teacher education programs to teachers' implicit beliefs and practical knowledge. Schoonmaker (2003) urges that teacher education move toward a deeper understanding of teachers' cognitive development, and to do that it would be first necessary to begin to address the implicit beliefs and practical knowledge they have constructed for themselves through their own sociocultural and historical experiences. But tacit knowledge is often situated within specific situations making it difficult to formalize and articulate, and this is one of the greatest difficulties of gaining understanding about teachers' beliefs and practices (Schon, 1983; Sternberg & Horvath, 1995). It is suggested further, that effective teachers tend to demonstrate enhanced self-understanding, and this facilitates the development of a positive, self-ethnic identity and self-inquiry into the relationships between fundamental values, attitudes, beliefs and their teaching practices (Gay, 1995; Sachs, 2004). Williams (1996) makes a strong argument by highlighting the fact that student teachers bring with them "implicit theories" in the form of cultural beliefs, traditions of childrearing, and anticipated student behaviors, which may very well clash with the "explicit theories" they are taught in teacher preparation programs. Further, although these "implicit theories" are very powerful and are, indeed, what shape a teacher's practice in the class room, the teacher educators tend to view them as barriers to good teaching. Williams suggests, instead, to view previously held assumptions as the first step that teachers take in beginning to build theory as the process of interaction between implicit and explicit theories is started, and reiterates the intimate connection between teachers and their values, suggesting that the teaching act itself is an actualization of a value system (Williams, 1994).

My research data revealed numerous examples that illustrated a clear process of sociocultural–historical construction that the teachers in India engaged in while learning and developing their teaching competencies; competencies being defined by that set of values, skills, and knowledge that would equip them to teach effectively in their classrooms. Their stories allowed me to gain an insight into their meaning-making processes and how they interpreted certain educational ideas.

Making Sense of Early Childhood Concepts

Charu, Malvika, and Vasudha used informal and nontheoretical language in describing how they implemented the educational aims in their classrooms, and how they made sense of terms such as "culture," "child-centered," and "culturally appropriate"—concepts frequently cited in the discourse of early childhood education in the Euro-American progressive tradition. For all three participants, formal teacher education had primarily prepared them to teach higher grades, and they had not been formally prepared in early childhood education theory. Starting with the word "culture," I asked them to define it within their own frame of reference.

Charu said that culture included "the basic norms, the basic values, the way we go about life. That is our culture. Like when I'm teaching my own children . . . I try to tell them these are the ways, this is how we led our life . . . these are the things our family believes in and these are the things I want you to learn and remember. Okay, [some] things do change but [not] the basic values and customs that our family follows . . ." This perception of culture as a set of underlying beliefs of a philosophy supported her practice as a teacher. Her goal to impart values to her students was evident in many of the examples provided in previous chapters. Charu recognized the importance of a culturally appropriate environment for children in the classrooms. When reflecting on the relationship of culture to education she referred to several American and British professional journals and magazines on early childhood education that were made available to the teachers in her school as reference and resource materials, and she explained how as teachers one had to modify what they read in order to implement it in their classrooms: "... in fact, they have so many magazines like *Childhood Education* ... A lot of their things are based on a different culture, things are very different from what we do. So you have to incorporate everything according to the background of these children, according to our own background. You have to mold it ... you have to keep changing it to suit your situation and environment ..."

Malvika's understanding of the word "culture" was that it was something that, although could be shared at a basic level, was manifested differently by different groups of people. To her, culture meant the following:

> morals, values, your code of conduct, your behavior, your way of seeing things, your way of accepting things, your way of dealing with things. That is what comes to me whenever I talk of culture ... we can say that culture is whatever the teacher has taught or the parents have taught ... there are certain mores, values, behaviors which are totally different for people. But still there are things at some basic level that all of them share. See, because whenever I greet my elders I would greet them in some way, and you may greet them in some other way. But all of us, whenever we meet our elders we greet them respectfully. So we may have our own ways of doing things, we may have our own values, but we are all doing the same thing.

Vasudha also explained culture in terms of norms and principles that are followed by a group or stratum of society. However, she delved deeper into her explanation and illustrated that even within the Indian culture there are subcultures:

> ... there are some principles and norms that are followed in every group and strata of society. So culture means ... principles and norms. Like, here [in India] we have a Punjabi culture, we have a South Indian culture [as two examples]. We have these principles, these norms, these different kinds of festivals, the different kinds of celebrations, the different ways of eating, different habits. So—everyone has a different culture. ... there's a different culture being followed in schools also. Like SD School will have it's own culture, and The School will have its own culture, and the Convent school has its own culture.

Vasudha shared an important understanding of the evolving nature of culture when she spoke of different parenting styles and said, "the culture of our parents was very protective [meaning overly protective of their children] ... don't talk much to boys and all. I have changed a lot. I would have no problems with my daughter's boyfriends coming over. Still I'm very Indian. I don't want her to go astray." Vasudha's understanding of culture seemed to mean, in fact, an Indian way of life and a set of norms in addition to descriptors of ethnic diversity that she has mentioned earlier. But also highlighted in the conversation was her recognition of the fact that culture evolves. She acknowledged that even though she shared the same culture as her parents, she had changed somewhat in her childrearing approach and her parenting style might now be different from that of her parents. Although Vasudha recognized different subcultures such as the Punjabi culture, the South Indian culture, and so forth, she also simultaneously acknowledged the element of Indian-ness underlying these different ethnic groups.

The teachers' responses with regard to the meaning of a "child-centered education" was also rooted in their own personal perceptions and understanding of this phrase. Charu perceived it as a program where there are very few children and where the teacher is able to give much more time and attention to each child, "where each child is not just listening but each child is interacting with the teacher . . . more of a one-to-one relationship with the child. I think so, but I'm not too sure. We cannot—in our situation—have that kind of a child-centered program with such large numbers [of students]."

Malvika's conceptual understanding of the term "child-centered" was in terms of the child's needs, which, as she explained, meant "deciding things based on what the child needs—with the changing environment, with the changing society . . . you know, the child's needs are also changing constantly . . . This is the way I define child-centered—something that the child needs, going according to the needs of the child, and then forming your curriculum on the same basis, especially the non-academic curriculum. . . ." Malvika admitted that the academic curriculum that was prescribed for all schools did not leave teachers with much leverage to proceed according to the needs of the children. She, in fact, referred to the given situation as teacher-centered education because teachers had to maintain a teaching pace that would allow them to complete a set curriculum within the allotted time. When I asked her about the role of the teacher, Malvika answered by saying, "she has to be a facilitator . . . I'm not trying to use advanced words, but to help the child learn about himself, and learn about his own abilities . . . teacher is also a major learner . . . it's not just teaching but it's a teaching–learning process. . . ." She made a point to carefully explain to me that the terms she was using were not just abstract theoretical terms but were actually quite meaningful to her.

While explaining her understanding of "child-centeredness," Vasudha attempted to take the perspective of the child and used her experience with her own daughter to explain her understanding of the phrase. She described it by saying:

> If I'm sitting on the other side, not as a teacher but as a child, how will I take it if a teacher scolds me? How will I take it if I'm not able to understand the

language she's using . . . So I always look at the other side, and since I have a small daughter myself, I always take the approach that if I would say something to my child, or if the child is not able to understand, if I would lose my patience . . . how would my daughter react if somebody else uses that kind of reprimand, or a kind of body language that the child will not like. So all that has to be kept in mind . . . also what is most important is the mood of the child. That has to be kept in consideration . . . The child has to be kept in mind.

Her sensitivity to being mindful of the child guided her perception of how children learn. Without referring to learning theories, she observed that the degree of children's engagement was an indication of whether they were learning and comprehending and enjoying the moment. In giving an example to illustrate what she meant, Vasudha recalled an occasion when she conducted the school assembly and based it on a story from the *Ramayana*. Initially she feared that this complex Indian epic story might be too abstract for her young students. However, she tried to make it as simple and interesting as possible while maintaining the authentic format. She seemed very pleased by the results of her efforts as she described what she had done: "I had written the lyrics also for their age level. I was so happy that I had written something on my own and something that the children could relate to and they enjoyed it . . . I had to put in a lot of effort in preparing them . . . But the show was really good and everyone enjoyed it. I could judge this because all the children who were watching the program were quiet and absorbed."

Throughout these discussions about early childhood education concepts, I found it very interesting that there was no mention of the names of any learning theories or any of the early childhood educators and theorists such as Dewey or Piaget even though Charu, Vasudha, and Malvika had studied them in their teacher education courses. The closest a teacher participant came to using a theoretical term was Malvika's mention of teacher as "facilitator," and in her eagerness to clarify that she was not using this term to appear theoretically knowledgeable, she inadvertently emphasized the divide that has been perceived between the theory and practice.

"What Influenced My Learning and Practice as a Teacher"

The general expectation that teachers will apply theoretical structures studied in their teacher education programs to the practical situation in their classrooms was definitely not being implemented in this study.

These teachers did not seem to draw upon the theoretical structures they had studied, which in their case largely reflected Western theories and philosophies of education, and instead were more inclined to draw upon personal beliefs and values during their classroom teaching. Further, on the basis of the interviews with Charu, Vasudha, and Malvika, as well as from other teachers' responses to the survey questionnaire, the fact emerged that the strongest and most defining influences on teachers' classroom practice came primarily from informal sources and to a limited extent from their formal teacher education. Informal experiences had a deeper and more lasting impact, and included teachers' personal experiences with their families and friends; their own schooling; and professional experiences such as workshops and seminars shared with colleagues at their workplace. Formal experiences had a more limited impact and included what the teachers learnt in their teacher education college during class lectures and student teaching. Vasudha, the Nursery class teacher, succinctly summarized this by saying, "The teacher learns every day . . . from her own students, from her own family, even from her own children, plus from what she has experienced being a student, and from the B.Ed. program also to some extent . . . plus all these workshops that are arranged for us." It was interesting and very telling to mark the order of sequence in which she mentioned these influences.

The nature of the various informal experiences that influenced the classroom practice of teachers, mainly those associated with their families, their own childhood teachers, their present colleagues at work, and also some of their formal learning experiences in their teacher education programs, are now taken up for discussion.

Learning from Family: Multiple and Lasting Influences

Charu, Vasudha, and Malvika all shared that their families had been very powerful influences on their learning to become teachers and their classroom teaching. Their educational philosophy and practice appeared to reflect their own personal values that had been instilled at home by their parents and other members of their extended families. Charu was very clear in stating that her father, even after his death, continued to be a source of inspiration for her in deciding what she taught her own children and the students in her classroom, especially her emphasis on environmental issues such as the conservation of water, electricity, and paper, and her emphasis on children being self-reliant and doing their personal work themselves rather than depending on servants and maids: "As I told you my father was a very wonderful person. And I always admire him, and I always look back, if I ever have a problem or I'm stuck somewhere, I try and think how he would do it. So that way he was a role model for me. So the things I learnt from him, I'm passing those things onto my children . . ."

As a researcher, it was interesting to note how each teacher described the manner in which children learned values from adults. In almost all instances, values were not "taught" but modeled by the adults in the lives of the children, whether they be parents, grandparents, or teachers. When asked specifically to describe how her parents taught her values, Charu seemed surprised I had asked her that, and she answered: "They never did. I just saw it, and around them their friends were like that, we were like that. I don't have any memories of them speaking to me about it or saying '*jhoot nahin bolo*' [do not tell lies] . . . I don't have any vivid memories of them preaching it to me. They might have. But my memories are of what they were [in their behavior]. They were like that . . . I know they would never have done anything wrong." Charu's all-inclusive and multicultural approach of celebrating all holidays and attempting to include even the poor employees of the school is also rooted in the way she was brought up. She said that her parents gave her and her siblings a very "broad" upbringing and "we never spoke of anybody as a South Indian, or a Punjabi, or a Gujarati. Never. For me, I just knew everybody as an Indian . . . My father would never be rude to a servant, or anyone who was working below him . . . That's what I learnt when I grew up." Charu admitted there were difficult issues also when she was growing up and recalled her relationship with her mother: "I . . . felt that she was my biggest enemy. She was quite strict with us . . . she wouldn't let us go out to parties and stay out late at night with friends . . . But I guess I've realized now that those things [rules] were really important." With her own daughter a teenager, Charu had started to believe that it was important for a parent or adult to be clear and strict about boundaries. As a teacher, Charu said that she was strict with the students in her class, but, at the same time, was also very loving. When I asked about the most cherished moments of her life, Charu answered without any hesitation: ". . . to be with my family, and my friends. Going out with my children. Renting a movie with them or going out to dinner with them, or whatever . . . You know—just being with family . . . I do most of the things with my mother . . . And specially for things like shopping—I like to go with my mother and my sister . . . Being with my family first, and then friends [are the most cherished moments]". This kind of family closeness and interdependency would seem strange to most youth and adults in the West, where shopping and watching movies are activities for which the company of friends and peers are preferred over parents and siblings. In addition to the importance of family, friendships have also featured prominently in Charu's life. She claimed to be a "friends person" and has always maintained friendships for a long time, without breaking them easily. The overlap of friends and family in her life was further clarified when she told me "my mother also has always had a very good set of friends. She also—when we were young—had four or five friends whose children are still friends of ours. And you know, we've grown up together and now whenever they're in town we meet and their children have got to know my children." Another influence Charu acknowledged had an important role was her own intuition and instincts in her practice as a teacher: "In fact, as a human being also I do things by intuition and instinct. I trust my instinct . . . It really plays a major role in my life and as a teacher."

Malvika voiced very similar thoughts about her own learning. She too spoke of the tremendous influence her parents had had on her classroom practice as a teacher, and said, "my mother is not a very strict person, but I'm very scared of her . . . she's my yardstick for measuring whether I'm doing wrong or right. Father is more like a friend to me . . ." It is interesting to note that she made a distinction between a parent who is strict and a parent who is more of a friend, and that it is the former from whom she learned more. The balance seems to have provided her with a very nurturing and happy environment because she went on to say: "Things have been cool for me. . . . I must say I've really enjoyed my childhood. And even today I'm loving every moment of my life." Malvika spoke of learning from her grandparents: "My grandparents used to live with us before they died. And you know, I used to love my grandparents more than my parents. And I still remember them. And I don't think I can ever love my parents as much as I used to love my grandparents . . . you learn a lot from grandparents. You know, parents have to teach you certain things. Grandparents don't teach you but then you actually learn so many things. So there's a 'learning without teaching' sort of thing going on with your grandparents". Malvika had grown up in a family that observed many traditions, the most important being the *Diwali Pooja* (the prayer ceremony to celebrate the religious festival of *Diwali*) that their family would not forego under any circumstances. Apart from holiday celebrations, several other family customs were observed: "we still follow whatever has been passed on to us by our grandparents. My mother is pretty strict in this regard. At times, let's say, we are going out . . . and we see a *khaali ghada* or a *khaali* pot [an empty clay pot] we think it is unlucky for us and we turn back and go some other time . . . we still light the *Diya* [a traditional oil lamp made of clay] in front of the *Tulsi* [the holy basil planted in the courtyards of many Indian homes] every evening . . ."

When I asked her to explain how her personal experiences were translated into her classroom practice, she was very clear about the close connection between the two: "I think that does get reflected . . . I pray every time before I eat a meal. So when the students open their *tiffins* [snack or lunch box brought from home] I tell them that you must pray to God and thank God before you eat. See, this is one of my personal practices which somehow I try to ... it's not that I try to impose it upon them. I don't punish them if they don't pray. But then I tell them to have this good habit to thank God for whatever he's given you—so it [personal upbringing] does get reflected everywhere." Being a young single woman, Malvika lived with her parents and her younger college-going brother and she recognized the close-knit relationship between the members of her family describing how "when Papa is back home [from work] in the evenings, we want to be together, we want to dine out together. We want to be together most of the time." Again that closeness and interdependency between family members was clearly evident in preferring to be in each other's company.

Vasudha reflected on her childhood experiences and the influences of her family, as well as the influence of her husband's family after she was married. She remembered being very social as a child, and was constantly in and out of neighbor's houses, climbing walls and trees, and playing sports. She observed how much simpler the pleasures of life were when she was growing up as compared to her daughter's activities today, which are so influenced by the West, such as skating, swimming, and elaborate birthday parties: "Now we celebrate more ... we spend a lot of time [in planning and organizing birthday parties] ... For her first birthday I worked for one and a half months and I called [invited] around two hundred people ... sometimes I feel why is it that my parents never celebrated in that way ... But even then the bonding was much greater at that time."

As part of her nursery classroom curriculum, Vasudha celebrated several religious and ethnic holidays, and had an India Week in which children in the class shared foods and clothes from different states of India, which she said was partially influenced by her husband's family who observes these celebrations at home in order to give the children in the family a broad and culturally inclusive upbringing:

We celebrate all the festivals here [at home] ... Like for *Janamashtmi* [*Lord Krishna*'s birthday, a national holiday] I dress up my daughter as *Krishen ji* [a respectful way of addressing *Krishna*] ... we'll all wear yellow clothes. Though my in-laws are not very religious. Specially my father-in-law, he never likes to visit a temple or anything ... But for our children we do everything—we dress them up, she'll wear the *morpankh* [a peacock feather like *Krishna* used to wear in his hair] and we'll decorate everything. Then for *Diwali* we'll make her a *koli* [a rural woman] and all. We'll decorate the house with the *torans* [ornamental and embroidered scalloped door hanging] and everything we do in a traditional way—giving of gifts, calling people over ... Then for Christmas we just made a Christmas tree out of paper ... We took the kids out to wherever something was going on for

Christmas—*Dilli Haat* [a local craft bazaar], and fairs, and all the hotel lobbies decorated with such lovely Christmas trees . . .

When I asked Vasudha about her most memorable moments, she too, without any hesitation, said that her most cherished moments were the times she spent with her grandparents. She described how attached she was to them, and shared some of her reminiscences with me:

> My *Daadi* [father's mother] used to sit with me. And though she used to be hard of hearing, I used to speak right into her ears. And there was not even a single day that I didn't speak to her . . . Whatever money I used to get for *Ashtami pooja* [a prayer ceremony] . . . I used to give it to my grandmother because I used to feel that was the safest place. And then I would go to my *Naani*'s [mother's mother] house . . . And I used to relish her food and never liked what my mother cooked. Always relished my *Daadi*'s food and whatever my *Naani* used to make. And I really miss my grandparents.

This intergenerational bonding was further facilitated by the fact that her grandparents' lives too revolved around her and her siblings. Vasudha noted that she was pleased that the cycle was continuing and that her daughter received the same amount of love from her grandparents: "I feel in my own house, my father-in-law and mother-in-law . . . for them—more than their own children—the grandchildren are most important." Vasudha's belief in the important place that grandparents have in children's lives was not restricted to her home environment. Supported by the school, she brought that special relationship between children and their grandparents into her classroom environment, and each year celebrated a Grandparents Day when only children's grandparents were invited to the classroom and not their parents. It was important, according to her, that every child had at least one grandparent attending. The grandparents themselves would be so thrilled at being invited that they would travel long distances from cities in other states, such as Ludhiana in Punjab, and Jaipur in Rajasthan, to attend this event in their grandchild's school. Vasudha's decision to become a teacher had been influenced to a great extent by members of her extended family: ". . . my own *Chachi* [wife of the father's younger brother] was a teacher, my *Tai* [wife of the father's older brother] was a teacher, my *Maamis* [wives of the mother's brothers] were teachers. And they were able to spend a lot of time with their own children. So maybe because of that I wanted to become a teacher."

Vasudha shared that her personality as a teacher had been shaped by her mother and also by her own daughter. With regard to the former, she recalled a day when she was a young school girl and was out walking with her mother: "Once a long, long, time ago on the street I had met a child from my own school and I just ignored her. My mother scolded me so much . . . I felt very bad and since then changed my attitude, even to the children in my class." More recently, after her own daughter was born, Vasudha had learnt how to become more patient with children because she felt that if she or someone else were to be impatient with her daughter, how would she as a mother feel? In other words, she was able to take another perspective after she entered motherhood. This had a direct impact on her educational philosophy and was evident when she spoke of the "Joy Quotient" as an educational objective, and that the first thing she wants for any child is that the child should be happy.

The deep influence on Charu, Vasudha, and Malvika of not only their immediate family but also that of the extended family was strikingly evident in several of the conversations. The kinship patterns that make up the Indian family system is structurally complex and each position has its own title, and the importance and authority of the title brings privileges. But as I have mentioned earlier in the book, with privileges also come responsibilities and duties. A child at home is thus influenced, taught, guided, and loved by all the women in the family as was the case in Vasudha's family, where all her various aunts influenced her decision to become a teacher. In short, the phenomenon of multiple mothering exists widely and the child grows up in a nurturing environment in which he or she is deeply loved and cherished.

Learning from Childhood Experiences in School

All three teachers acknowledged that their school experiences and their encounters with teachers also had a strong influence on their teaching styles. The experiences and encounters described were both positive and negative.

During my first conversation with Charu, she mentioned that she had attended several schools during her childhood years because her father worked in the government services and during his career had been frequently transferred to different towns and cities in India. On reflecting back, Charu was able to connect some of her teaching methods to the styles in which her teachers had taught her. In fact, she realized even while speaking with me that the way in which she taught English mirrored to a great extent the teaching style of her English teacher: "Whenever I'm doing a lesson I do it the way I learnt it, you know. Even things like words—when I'm explaining a word. Never just the meaning. We first act it out . . . That was something I had learnt when I was in school." In particular, she highlighted her preference for sharing personal anecdotes and experiences from her life with her students, and noted that they were far more attentive when she did that than when she taught them a lesson in a more formal manner: "I can remember a few teachers I really admired . . . Probably because one of them told me things like this . . . She was our class teacher, and in the morning we had a little time with our class teacher . . . And she used to tell us lots of interesting stories and anecdotes and experiences. . . ." As she reflected on her educational philosophy, she recognized that her style as a teacher was also influenced on an informal and subconscious level: "Probably I don't even realize it but a lot of it must have come from my school days also. I mean I don't have a conscious memory of it but I'm sure there must have been things like that. Which is why I feel so strongly about certain things now."

When she was in Class Six (sixth grade) Charu had to go to boarding school because there was no good school in the small town where her father was posted at that time. I found it quite fascinating to hear her recall and describe the regimental life in her school hostel where "everything was on time. You had to study on time, eat on time. Eat what was served to you. Not like the way our kids behave. Things were really strict. We used to grudge it at that time. But now when I look back I feel that [the strictness] was really very important for me to grow up as a person." This led her to reminisce about her experiences at home with her parents and she recalled that her mother had also been strict with her. On hindsight, she felt that in both cases, the strictness of her teachers and her mother had been good for her and had made her into a better person as an adult. Although initially she had a traumatic couple of years at the boarding school, Charu still thought of her school experiences with fondness, and remembered it as being great after the initial two years. She had started to enjoy all her school activities as she became used to it and began to feel increasingly comfortable in the boarding school: "And I remember that was very traumatic and for a year or two I just couldn't take it. Though I had two cousin sisters who were also studying there. And I used to really cry. It was probably the separation from my mother. Every time I would get a letter from her I'd sit and cry . . . I remember those were two years that I was very unhappy. But after that it was really really [emphasis] great. Probably I grew up and made friends and started enjoying all the activities that I had there . . ."

She described her teachers as being very dedicated: "We were a smaller number of children, and they knew us very well. And they were really dedicated. Completely. We had a lovely experience. I mean, after my initial first two years." Charu seems to have emerged with a sense of hope and optimism after having been through some hard times as a child—with the belief that hard times are a fact of life and that the test of good character is how you conduct yourself during moments of hardship. Her current classroom practice is based on this same belief that clear and firm boundaries are critical along with plenty of love and support, and this balanced teaching approach helps children grow into stronger human beings.

My conversations with Malvika, on the other hand, revealed a sociocultural tension as she admitted to a distinct disparity between her school life and home life, which led to her actually manifesting different personalities "one at school and the other at home. The moment we used to get off at our bus stops we were a changed personality." This, according to her, was largely due to the fact that the school she attended focused primarily on academic achievement and did not provide an environment within which could be acknowledged the sociocultural background of its students. In a touching admission, Malvika said: ". . . I was actually scared to share my feelings, my values at the school because I was scared to be the laughing stock of the school. If I say something—about my rituals or something—everybody, including my teachers are going to laugh at me." Her words indicated that there seemed to be an implicit expectation in the teaching profession whereby teachers were not supposed to combine the sociocultural lives of students with their academic lives. This seemed to be a direct reflection of how the teacher education programs had been preparing teachers, and in many places continue to prepare student teachers with the goal of teaching them the technical skills required for successful academic teaching. Malvika recalled feeling that she ". . . really loved those teachers who were something more than teachers. Not teachers who only did the academic part, but something more than that. . . ." Perhaps that experience is the reason why Malvika herself strongly advocated the teaching of Indian values in the classroom, and recognized that it was critical for the teacher to acknowledge and respect who the child is at home and his sociocultural context "because that's who the child really is." She described a particular experience from her school days:

> . . . there was a physical education teacher and certain students used to touch his feet because he was very caring and loving and he used to bless them every time they used to touch his feet. So one classmate who came from a very high [socioeconomic] class, even he used to touch the teacher's feet and he used to enjoy the blessings that the sir used to give him . . . he started imitating others [other students] and began to do something that was never done in his own family . . . And I think he must have carried it on his life. I mean, I don't know where he is now. But wherever he might be he must be touching people's feet just to show that I respect you, I care for you, I love you and all that.

The class issue emerged several times during Malvika's interviews. In grappling with her own feelings, she shed light on how students from a lower socioeconomic class might experience the social class tensions in a negative way when they attend a private school where the majority of students belong to higher socioeconomic backgrounds. In her stories was also embedded her implication that traditions and values were practiced more actively in working middle-class families rather than in richer middle-class families. The other interesting perspective was her understanding of the act of touching a teacher's feet as being a blessing and an expression of love and respect, rather than a negative experience. From a Western egalitarian point of view where feelings of love and care are verbally communicated between individuals, this act could be seen as one of subservience. But again, one needs to be reminded that expressions of love and behaviors that communicate care and respect can take different forms in different societies. While growing up in India, I don't remember a single instance when my parents or grandparents verbally told me that they loved me. But I knew from their expressions, gestures, tone of voice that they did deeply love me, that I was of the utmost importance to them, and that they would have done anything to ensure my happiness and protection. The same is true of my relationship with my sons. Amongst people in India, words are not quite as important as the actions taken in demonstrating love, responsibility, and caring to immediate and extended family members and friends.

Vasudha held very fond memories of her childhood and school days and it was quite apparent that she had had positive school experiences. During her reflections on the strong family bonds in Indian society described earlier in this chapter, she added, "when I was a child, that bonding was there even with some of our teachers. I still feel very much for my teachers, the way they loved us, the way they cared for us. . . . " Vasudha maintained a happy relationship with not only her teachers but also with her classmates with whom she still kept in touch, and described her school as being fun and a place where she enjoyed her days so much that she just wants "to go back to that school." That experience shaped her own educational philosophy whereby the most important goal for her is that the child be happy and that early childhood teachers should strive to develop the joy quotient in their classrooms. However, although Vasudha had an overall happy experience her school days were not without painful moments. In her senior year, she felt devastated upon hearing that she was not going to be presented with a highly coveted award for winning the maximum number of gold medals in the area of sports, and instead, the honor was going to be awarded to another girl. Vasudha's pain was revealed as she described that incident to me:

In Class Twelve [twelfth grade] there is an award ceremony and only the *Twelfthees* [high school seniors] get those awards. I had participated in lots of sport activities. And I won many gold medals . . . There was another friend who had also won . . . So when it came to the main award for the Annual

Day, I came to know through somebody that I wasn't being given the award. It was going to the other person. And I cried and cried and I howled. What I felt was that I also deserved it . . . In fact, if you looked at the competition, she [the other student] was only competing with the children of our own school. I was competing with the whole zone and I had won gold medals in that . . . I really cried a lot. All my class teachers, subject teachers and all my sports teachers fought with the Principal and in the end both of us got the award.

Clearly this experience left a deep impression on Vasudha and influenced her current practice as a teacher. In her personal educational philosophy, Vasudha affirmed inclusion and the fact that all children should be included in classroom activities and projects. She said that her goal was to ". . . encourage one hundred percept participation in all the school activities. Normally whatever functions we have, whatever assembly programs we have, we have a hundred percent participation. All the children are involved . . . our aim is that no child should be left out, nobody's heart should be broken."

Learning on the Job and from Collegiate Experiences

All three teachers maintained that since their teacher education program prepared them to teach only higher grades, most of what they have learnt as early childhood teachers had been from their encounters with other teachers in The School, workshops and seminars offered by The School, and mentoring and advisement provided by their supervisors.

Charu described that before she started teaching early childhood at The School, she was a geography teacher for Class Six, Seven, and Eight (sixth, seventh, and eighth grades) in another private school in New Delhi. She was thus hired with no formal experience and training in early childhood education. She remembered how helpful she had found one of the workshops she attended: "Specially for teachers who recently joined, initially for the first few years they are sent for—workshops. I'll never forget—I'd been to this workshop—this Math workshop . . . Such a tremendous impact it had upon me, as a Prep [kindergarten] teacher!" Charu referred to one of her supervisors, Mrs. Kumar, as someone who guided her when she first started to teach at the early childhood level, and how much she valued what she had learnt from her:

Like, we had Mrs. Kumar—she was in charge of Prep, One and Two [kinder-garten and first and second grades] . . . She was really good. She was very strict, but she would guide you very well . . . and most teachers hadn't done what they are currently doing before they came here—I'd never taught Prep before I came here. So how to go about the lesson, and how to read poems and how to do sounds and formation of letters . . . we had never learnt that in our school. I didn't know all that when I came here . . .

Guidance and help was also forthcoming from other fellow teachers at each grade level. Teachers took turns to write the lesson plans every month and shared it with their colleagues. This led to less disparity between the curricula in the different classrooms, while increasing dialogue and interaction between teachers. Each grade level had a Rep (a grade representative) who was one of the class teachers, but who also had the responsibility of coordinating all the classes at that grade level and who would be sure to discuss the curriculum with all the teachers so that everybody was kept informed and was on the same page.

Vasudha described a supportive school environment that was conducive to learning the necessary skills to be an early childhood teacher through many workshops and seminars. However, she also noted that her own instincts played a large role in shaping her teaching methods, more so than the seminars. Mentoring by Veera, who coordinated the nursery classes took place on a regular basis in the form of discussions on curricular matters, appreciation of good practice, and scheduling of workshops. "We exchange ideas with each other—we're a very close knit family," Vasudha said of the nursery class teaching community in The School. She spoke particularly of how much she had learnt from a previous headmistress:

> I've learnt a lot from her. She was a very strict administrator, a strict headmistress, *and* a loving and a caring one also. She has a reputation in the other school that she is a very strict teacher, a strict disciplinarian, but I have always loved her a lot, and *now* we've all realized that it is all because of her that we've become, umm, good teachers. Like, [earlier] I couldn't write, I couldn't create anything—she used to give us so much scope to write, to think, and what I am today is because of her.

Vasudha concluded that although teachers learn a lot from professional workshops, their families, and also from their teacher education programs, the teacher learns everyday and continues to learn: ". . . basically the teacher learns from the classroom. Maximum from the classroom. She knows the needs of the children . . . and every time there's a different group, the group has different needs—so the teacher learns from the child."

Malvika had not as yet had an opportunity for attending professional development experiences organized by The School since she was still in her first year of teaching. However, she did have a coteacher to work with who had been there longer than Malvika herself, and who was a big help in offering guidance and advice. Both Antara and Veera, the immediate supervisors of the three teachers, agreed on the fact that they were available to speak formally and informally with any teacher, and sit in on discussions and planning meetings with them, in addition to facilitating numerous workshops, interactive sessions, and teacher education seminars for their staff.

Learning from Formal Teacher Training

Upon reflecting on their growth and learning as teachers, all the three teachers, in separate interviews concurred that their formal teacher education programs prepared them mainly in technical skills that equipped them to address the teaching of academics. Their capabilities and skills for teaching values and building character in the children grew out of their personal experiences. Judging from the responses to the surveys and interviews it appeared that some of the skills that teachers learnt during teacher education and which they found to be helpful were learning how to write lesson plans, learning how to introduce a lesson and teach it in units, learning how to manage a class with large numbers, actual classroom experience through their student teaching sessions, and technical courses such as "Audio-Visual Aids" and "Teaching Aids." Most of the theoretical courses on child psychology, philosophy of education, history of education, and so forth, were found to be boring and remained disconnected from their practice. The teachers admitted that they studied many theories in the above courses but most no longer remembered the names of any theories or theorists now.

Charu seemed strongly critical about her teacher education program. From the very beginning of her program she felt the institution to be a cold and impersonal place where there was "nothing to make you feel that you belong . . . and it was like being lost in an unknown place." The classes were large university classes with no interaction between the students or between students and the professors. The classes were "theory oriented . . . most of the lectures made no sense . . . we learnt them because we had to put them down on paper." Even though she felt learning how to write lesson plans was helpful, Charu added that they were merely given a formal format and there was little guidance or support from professors if a student teacher wanted to make some innovations or try out a new approach in teaching a lesson. She recalled an elective course called Audio-Visual Aids, which helped her tremendously during the time she taught Geography in middle school. The practicality of a teacher education class or course seemed to largely determine whether the course content stayed with the teachers once they were in their own classrooms.

With regard to the issue of whether she felt she was prepared to teach the children the values, skills, and knowledge that she held to be important, Charu said that only came with her experience. The teacher education coursework did not touch upon the teacher's role in imparting values:

> These things were never touched upon. I'm doing it because I feel strongly about it . . . So these things, I feel, during our teacher training should be incorporated into the syllabus. That this is the role you play as a teacher. That should be a very important part of the teacher training. And what your responsibilities are as a teacher. How important a role you play in the life of children . . .

The underlying implication of this response seemed to be that just as early childhood classrooms aimed to develop the whole child, so also the teacher education programs should aim to develop the whole teacher and prepare her to teach the values, skills, and knowledge that are prioritized. Charu's B.Ed. degree had prepared her to teach a content area (Geography) in grades six through twelve, and she felt very unprepared to teach an early childhood classroom. She reached a comfort level for early childhood only after she had had a couple of years teaching it while she learnt from her own experiences in the school.

Vasudha was a Nursery class teacher but had a B.Ed. degree in teaching Science and Home Science to higher grades. Her student teaching took place in grades six through eight. She acknowledged several aspects of her teacher education that she found to be helpful and appreciated the several seminars, activities, lectures, and research work that they had engaged in as student teachers. She also appreciated the time and effort that her teachers and supervisors put into helping her learn how to make lesson plans that incorporated plenty of activities. She mentioned a specific professor who was very strict but thorough in her work, and it was the high expectations of that professor which provided Vasudha with the most meaningful learning experiences: "She would never be [easily] satisfied with what we would do. She would say go and study further and find out more materials . . . and she would make sure we would read . . . Really I learnt a lot. I became so independent. I started visiting libraries." Vasudha mentioned an elective course titled "Teaching Aids" that she found to be very useful because it taught her to make materials such as flannel boards, charts, flash cards, and big books. She still made these aids to facilitate her teaching in her current early childhood classroom.

However, I found it interesting to hear Vasudha's admission that she did not require the B.Ed. training in order to be an early childhood teacher. According to her, much of her teaching was rooted in her own experiences and instincts. On the other hand, she reflected that if she had to teach older students then definitely she would have needed the B.Ed. training for knowing how to teach academics and knowing how to employ appropriate methods and approaches in classroom management:

> ... when I came as a Nursery teacher here I didn't have any experience and I didn't know what to do and how to start it ... as a nursery teacher I've learnt so much that I hadn't learnt in my B.Ed. [In the B.Ed. program there were things] like discipline, reforming the children, don't use reprimands. . . . But I can't correlate that too much to what I do here [in the nursery class] because I'm teaching little children, and with little children you have to be *extremely* [emphasis] patient and *extremely* sensitive. I know even with the older ones you have to be extremely sensitive, but then you have to give them a little bit independence also. Here you have to be vigilant all the time, *all the time.*

Malvika had been a Psychology major before earning her B.Ed. degree, and I could see the influence of Western psychology and child development reflected in her choice of words, such as effort, initiative, self-identity, ego, and so forth, which are commonly used Freudian and Eriksonian terms. She was currently a first grade teacher although she had received her B.Ed. degree in teaching Science and Home Science to senior classes. Her student teaching had been done at the grade eleven level. Since she herself was a novice teacher and a recent graduate, her teacher education course material appeared to be still fresh in her mind and during the course of our conversations she did mention learning about the theory of Piaget and the philosophy of Rousseau. She, however, lamented that the student teachers were never given a chance to relate these theories to practice and all they did was to memorize the theory and reproduce them for the B.Ed. exams. The focus of the B.Ed. curriculum was only on academic teaching:

> How to teach them [students] seasons, how to teach them this, how to teach them that. And hardly any stress is laid on value education or on explaining our culture, or telling them about your culture. I think this curriculum also needs to be a little reformed . . . See, training [B.Ed] is one thing—you're being told how to handle forty or fifty students in a class . . . it never talked about the "sixth sense" or these instincts [in teaching]. Never. . . . B.Ed. is not only handling the academic part of school—but handling everything a child has.

It was evident that Malvika, who viewed education as working with the whole child felt unprepared as a teacher in the most complete sense after graduating from her teacher education college.

The Nature of Student Teaching Experiences

Paradoxically, each of the three teachers found student teaching to be the most interesting component of teacher education. Although the student teaching phase was very brief, and they faced many tensions during that time, it brought them into direct contact with real students and gave them a more tangible sense of being a teacher. Charu remembered "the best part was the practice teaching" because that made more "sense" as compared to the "boring" lectures on theories, which "mostly focused on . . . theoretical things of education—their education acts and all their policies which . . . we should know broadly, but we're not concerned with that in our day-to-day teaching." But even the preparation for practice teaching did not allow for any discussions on what the student teachers might encounter in terms of the socio-emotional dynamics of any classroom.

Vasudha had a comfortable experience with student teaching as she had been placed in the same school she herself had graduated from. Much of the curriculum was still the same and many members of the staff were familiar faces to her. She explained that the practice teaching, or student teaching block, occurred in two phases during the B.Ed. degree program, with 15 days each of teaching lessons in a classroom in the first and second terms.

Malvika expressed mixed emotions as she recalled her student teaching experiences and said: "I really really enjoyed my practice teaching, in spite of all the stress that we got from the supervisors and teachers. I really enjoyed it . . . that was the first teaching experience, with your new professional identity. . . . " She remembered moments that were acutely distressing to her with regard to the rigid expectations that student teachers were held to. The first was in terms of having to dress "appropriately" while teaching in the classroom, and Malvika described this traumatic experience when she had to teach a lesson in Class Eleven (eleventh grade) for the very first time:

> I had my practice teaching at ABC School . . . in Science and Home Science. It's a co-ed school. The principal on the first day told me that you have to wear a *sari* [an 18 feet long piece of fabric that is draped and pleated around the body and worn by most women in India] and you have to put on a *bindi* [a colored dot that adorns the forehead of a woman in India] because you appear too young . . . It was a very fresh experience of wearing a *sari* for me and I wasn't very comfortable with it. So I had gone to teach Class Eleven and while I was walking between the rows [of desks] a boy stood up and said "Ma'am what are you wearing?" I was so scared. I thought that . . . my *sari* had come undone . . . I was so scared that I can never forget that experience . . . I was almost on the verge of crying . . . And then on another

day—my *sari* did come undone. I was in Class Seven [seventh grade] and I was dictating something to the students. Suddenly I stepped on the pleats of my *sari* and all the pleats came out. And then I didn't know what to do next . . . I can't forget that situation . . . But thanks to God, I was able to handle it.

Another incident that disturbed her occurred during one of the lessons that she was teaching. Her supervisor was at the back of the classroom observing her:

I remember . . . when I was doing my teacher training, my supervisor was sitting at the back and a child stood up and said "Ma'am, I've got a problem." I still feel sorry . . . I mean I didn't listen to him. I just said "please sit down, my supervisor is sitting at the back." If he stood [and interrupted the lesson] . . . then I [was afraid] that I would have bad remarks entered into my diary [portfolio]. I just didn't listen to him. I made him sit down . . . I could not ask him what his problem was just because my supervisor was sitting there.

Student teachers seemed to be assessed to a large extent on the way they delivered a lesson, and Malvika strongly disagreed with that practice saying that teaching was not just the delivery of an academic lesson, but involved emotional and psychological interactions with the children, which were largely ignored and disregarded by the evaluators. She regretted that she was not able to help that child who stood up during her lesson saying that he had a problem, and believed that her supervisor would have seen that as a disrupted lesson and most likely would have given her a low grade. The third concern experienced by Malvika as a student teacher was the tension caused by the difference in expectations between the supervisor and the classroom teacher:

. . . it wasn't easy handling the supervisors and the teachers . . . private schools have very high pressure and they [the teachers] are answerable for everything and anything they do. So they made us [student teachers] answerable to them in turn . . . they used to tell us—"this is what you're supposed to cover, this is how you're supposed to do it." On the other hand the supervisor used to pressurize us that this is not the approach you should use. See, we were at a crossroad—which approach to take. But the supervisor would never realize this is a problem . . . if we've asked for her [a teacher's] class specifically we need to give her our best and whatever she wants. And not what the supervisor wants. It wasn't that easy, it was quite difficult for us . . . it was a period of stress and strain, I must say.

However, Malvika reiterated that in spite of these tensions she enjoyed her student teaching phases when she really got a sense of being a teacher.

Conversations with the teacher educators Ashwini and Nagma revealed information that supported the perceptions of the teachers. Ashwini said quite clearly that most teacher training institutions in India offered a B.Ed. degree program with large class sizes, each having about 60 potential teachers, and the emphasis was largely on lesson planning and classroom management. She also criticized the sporadic student teaching experience of about three weeks as being very limited and not an ongoing routine as part of the degree program. Nagma acknowledged that most departments or institutes for teacher education in India deal with teacher education only for secondary school levels, whereas teacher education for primary and nursery schools is taken up by other institutions that are not affiliated with any university and which offer only diploma or certificate courses rather than degree programs. In fact, both Ashwini and Nagma explained that early childhood education as a field in India has not been included in the area of education but falls under the umbrella of the psychology-driven field of child development. They both felt that the curriculum for teacher education was driven by psychology, which tended to ignore the sociocultural dimension of teaching and learning. Ashwini specified that the "richness of our own culture doesn't come through because (a) our training is in the English language, and (b) the curriculum [for B.Ed] was designed during colonial times." This supports the concern that the teachers voiced about feeling unprepared to teach their students the cultural values component of education, in which case they have to rely totally on their own instincts and upbringing. Ashwini suggested that teacher training should offer a balance of formal as well as informal methods of teaching, pointing out that informal modes naturally begin to include "oral traditions and the funds of knowledge contained within these oral traditions known in Hindi as *Lok Vidya* [popular knowledge] that is found in local stories and conversations." According to her, this would not only empower teachers culturally but also provide them familiarity with the "child's inner world and social background."

A third aim of education that had been expressed by many of the educators, apart from the teaching of values and the teaching of academic content, was intellectual development and developing the ability to think independently. This has been described in detail in chapter 4. I bring it up again as it was interesting to note the distinction made between academic learning and intellectual development as two separate aims of education. Nagma supported the concept of having teachers as well as children become independent thinkers and was favorably inclined toward the model of teachers being reflective practitioners. However, she indicated that teacher education programs do not foster that model. Even though a handful of colleges in India were initiating experimental programs that encourage their student teachers to reflect on their learning experiences, Nagma found it problematic that when these teachers went out into the real world of schools, they had to surrender to the rigid pressures of the prescribed curricula and teach to the textbook. Thus the teachers' own practice of not being allowed to reflect on their teaching seemed to be mirrored in their expectations of their students too, and the importance given to reflective practices rapidly diminishes in comparison to the importance given to scoring high in tests and examinations.

Findings from the Survey

Similar perceptions about influences on learning to be a teacher were obtained when the wider sample of teachers participating in the survey were asked to respond to the following question:

Indicate which have been the two strongest influences on your learning to be a teacher (rate them 1 and 2):

Formal teacher education curriculum (B.Ed)
Your own teachers when you were in school
Your parents and extended family
Your friends and other social factors
Informal workshops, seminars, conferences at your place of work
Other teachers, colleagues and mentors at your place of work

Of the 32 teachers in the 15 different schools who responded to this question, 41 percent indicated that their family's role was most important; 25 percent indicated that their own teachers were most influential; 19 percent indicated that they learnt how to become teachers mostly on the job; 9 percent rated social factors as being most influential; and only 3 percent of the teachers indicated their formal training had the most influence on their learning to become teachers.

It is also interesting to note that a contradiction emerges between the image of a good teacher in India that has been established based on the teachers' own perceptions as discussed in chapter 5 and the skills that the participants learned most during their formal teacher education. The good teacher was described extensively in terms of dispositional and affective qualities whereas the skills learned during teacher education were more technical in nature. My conclusion regarding this contradiction is based on the information I obtained through my interviews with the educators as well as from the responses to the survey questionnaire I had sent out to the larger sample of teachers in New Delhi. The question included on the survey that addressed this issue was:

a) Do you feel that the B.Ed. course prepared you appropriately to teach your current class? If no, explain in what ways it did not.
b) List the skills and knowledge that you learned during B.Ed. which have been most helpful to you in your classroom work today.

Fifty percent of the teachers wrote that their B.Ed. course had not prepared them at all to teach early childhood classes from nursery through second grade, while the others did not indicate anything or wrote that it was only somewhat helpful. In response to the second part of the question, teachers identified the following kinds of skills and knowledge as what they had learnt from their teacher education (B.Ed.) program. The number in parenthesis indicates the frequency with which each response was cited:

- Lesson Planning (12)
- Teaching methodologies (9)
- Making and using teaching aids and audiovisual aids (8)
- Class control and time management (7)
- Child psychology (7)
- Blackboard writing skills (3)
- Arts, crafts, and painting (3)
- Maintaining teachers' registers and logs (2)

Interestingly, six of the teachers specifically mentioned that being a teacher depended on the individual's own personality and that they had learnt how to be a teacher from their own experiences inside and outside the classrooms rather than from the course work included in their B.Ed. programs. These responses were in consonance not only with what the three primary teacher participants, Charu, Vasudha, and Malvika, shared they had learnt during their teacher education experiences, but also with the research literature on teacher education in India, which suggested that the B.Ed. courses are tailored to prepare teachers in technical skills and methodologies rather than in dispositional skills required for the act of teaching.

With regard to the nature of teacher education based upon the findings of this study it is therefore reasonable to conclude:

- there was little recognition of who the teachers were as individuals, and how their socially and culturally constructed implicit belief systems came into play in their classrooms;
- student teachers seemed to experience a high level of stress due to the different approaches of their cooperating classroom teachers and their college supervisors;

- formal teacher education was geared more toward preparing teachers to develop academic proficiency in older students and less on teaching the whole child;
- teachers felt that there was no discussion on the moral responsibilities of the teacher during the course of their teacher education program;
- the environment in some teacher education colleges lacked in warmth, personal attention, collaborative group work, providing space for dialogue and discussion; and
- there was far too much stress on Western child development theory, which most teachers found to be unrelated to their classroom situations.

Two Parallel Processes for Learning to Teach

A significant understanding that emerged from the interviews with the teacher participants was that their teaching practice had two parallel dimensions to it as is seen in figure 7.1. One dimension was driven by the systemic focus on academic performance. This dimension assumed importance due to the prevailing system of education in India and the evaluation of students through a text-based examination system at every grade level. The second dimension was driven by the primary aim of the practitioners themselves—the teachers and school principals—which was the teaching of values to children and helping them develop into strong human beings intellectually, socially, and emotionally. Thus the two aims were manifested in two different forms of practice. The academic teaching showed up in a more concrete form by way of the formally prescribed curriculum through written assignments, prescribed syllabi, and textbooks. Values teaching showed up in a more verbal and informal form by way of a powerful variation of the hidden curriculum.

Teaching is about interpersonal interactions as much as instructional activity. On reflecting on their growth and learning as teachers Charu, Vasudha, and Malvika concurred that their formal teacher education programs prepared them only in technical skills that equipped them to achieve the teaching of academics, which in their opinion is the secondary aim of education. Their abilities and skills for teaching values and building character in the children grew out of their personal experiences in a sociocultural context. Sociocultural theories facilitate the understanding of how context influences the development of attitudes and beliefs of teachers. The findings could be supported by social cognitive theory whereby a powerful relationship exists among personal, behavioral, and environmental factors (Bandura, 1997). The findings also clearly indicate a circumstance that can

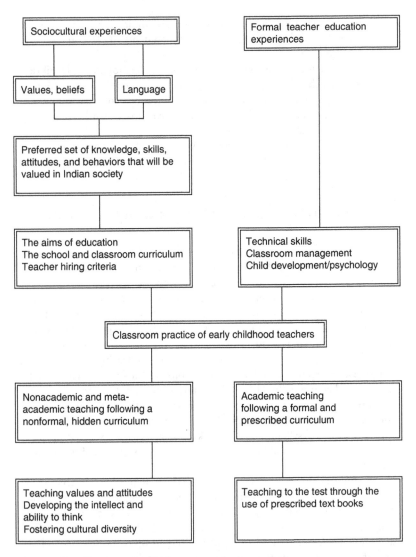

Figure 7.1　Two streams of influences on learning to teach

be viewed through the lens of the sociohistorical psychological processes Vygotsky described according to which community influence, interpersonal interaction, and intrapersonal reflection guides the individual's mental development (Sachs, 2004). With reference to the three teachers in my study, in spite of the fact that some formal learning through the teacher education colleges was helpful, most of the significant learning experienced by them was sociocultural in nature and resulted from internalization of external influences within the areas of (1) home, which involves social and cultural learning through interactions with parents, extended family, and friends; (2) school, which refers to the learning experiences teachers had in their own school and from the way their own teachers taught them; and (3) workplace, which provided mentoring and guidance from senior teachers, workshops, seminars, and staff discussions. This was also in agreement with the research literature discussed earlier in the chapter. The teachers' voices lend strong support for the perspective that learning, development, and growth happens on a social level first and then at the individual level after being taken into the classrooms by teachers. Sociocultural encounters faced by teachers during their childhood years are later internalized into personal theories of education during their teacher training and classroom practice moments. This phenomenon reflects Vygotsky's explanation of the same, as well as research by Stott and Bowman (1996), which suggests that values and beliefs underlie all theory and practice and to avoid a discrepancy between the latter two there needs to be consistency between personal beliefs and theory and practice. A more detailed discussion on achieving such a collaboration may be found in chapter 10 of this book.

Chapter 8

Contextualizing and Demystifying the Challenges of Large Class-Size in India

Charu's second grade classroom was furnished with only tables and chairs. There were 20 tables that seated 2 children each, and 40 chairs. The walls of the classroom were lined with bulletin boards that displayed several colorful charts and visuals. One bulletin board was prominently titled *Our National Symbols* and displayed illustrations of the symbols of India such as the national flower (lotus), the national bird (peacock), the national animal (tiger), the national fruit (mango), the national tree (banyan), the national flag, and so forth. Another bulletin board displayed a visual presentation titled *The Mahatma* (referring to Mahatma Gandhi). I noticed on the blackboard that the lesson that had just been completed prior to my arrival was called *Sources of Water* and there were questions listed on the board for the children to answer in their notebooks. It seemed that the thematic unit for the month was Water, and the students had been busy with the writing assignment. Charu was grading notebooks at the teacher's desk. As I entered, she asked the children to wish me, and all the children stood up, saying "good morning, Ma'am." The class was just completing a period of written work and as each student finished the assigned work, he/she would stand up to submit their notebooks placing them in a neat pile on the window sill.

Charu invited me to sit at the teacher's desk but I asked if I could observe from the side, asking for a small chair. "Ma'am, I'll get it," several students chorused. An extra chair was located and they rushed enthusiastically to bring it to where I was standing. "Please take out your Onward English books [the English reader]," Charu announced. One student asked for permission to drink water and Charu took the opportunity to announce a water break before starting the lesson. "Everyone, drink a little water," she encouraged. The students took turns to walk up to the windowsill on which were placed their individual water bottles in a row.

"Okay, water break is over now. Turn to page sixty-eight of your books," Charu directed. The lesson they were beginning was called The Ghost in the Bathroom. "What is a ghost? Has anyone seen a ghost?" As Charu began the lesson several hands went up and each student was given a moment to tell his/her story.

"I saw a ghost in my *Naani*'s [mother's mother] house. It looked like *dhuan* [smoke]," said one girl.

Another seven-year-old described how to talk to a ghost: "you call a ghost. Take a *katori* [bowl] and put a coin in it. Then take a pencil and put it on a piece of paper. Don't hold it."

Yet another girl said: "Once I went to a party and in the middle of the way there was a forest. A ghost came up and he was without a face."

I listened with fascination to the richly imaginative and graphic descriptions of these second graders' prior "experiences" with ghosts as they continued to get a chance to share their understandings and perceptions of what they knew about ghosts. All along Charu gently continued to suggest to them that there was in fact no ghost, and that the story in the book was all make believe. Animated conversation, talk and discussion continued for several minutes. Then Charu announced: "Okay, now no more discussion. Let's get back to the book. Perhaps you want to write a story about ghosts for home work?" "No! Yes!" the students responded excitedly, all answering at the same time. The atmosphere in the classroom was interesting and lively with a constant loud buzz. Charu proceeded to read from the book and during the course of her reading she would pause at certain words to explain what they meant.

"What is the meaning of lively?" she asked the whole class. Students raised their hands and she gave them a chance to answer.

"Almost always busy."

"Active."

"Very good," Charu responded. "People or children who are always active. Who do you think can be called lively in this classroom?"

"Ma'am, you. Because you're always teaching and running around," answered one boy.

"Name someone in this class beside me," said Charu, and immediately students began calling out the names of various fellow students.

Charu read another sentence and stopped at the word "hill town." "We usually call it a hill station. Who can name a hill station?"

"Simla, Musoorie, Nainital, Bombay . . . ," the students answered.

"No, Bombay is not a hill station," Charu corrected.

And so the class continued. The students were eager to answer although I did observe that the same 15 to 20 excited hands were always raised. But despite the fact that some of the other students were quieter, they were still very attentive to the discussion. One boy sitting close to me was cleaning his pencil-lined textbook with an eraser. However, it seemed that whenever Charu spoke, most of the students were listening attentively. She made the lesson interesting by connecting words and stories to the children's lives.

Talking about a *dhobi*, she translated the word into "washerman," and went on to make a connection to washing machines commonly seen these days. She gave them examples from her own life when she was a young girl in school, and often provided Hindi meanings and explanations for many of the words that appeared in the text. At one point in the story, a character by the name of Mr. Ali visited the house of some children and was welcomed by the children's mother. Charu stopped reading to reinforce the concept of hospitality and the importance of being welcoming to guests. "You should always greet people when they visit your house and offer them a glass of water or tea," she explained seriously and earnestly.

Charu read on, and used dramatic play to explain certain words and phrases. For instance, for the word "untied," she took off her *dupatta* (long scarf) to tie it around a chair and then untie it. Students volunteered to act out the word and a small scene was enacted. Dramatic play was used often to facilitate the understanding of nouns and verbs that appeared in the text. Students seemed accepting of the fact that only a few of them would get a turn to act. They were all eager to participate, but they also seemed to possess a basic understanding that not every one could get a turn that morning. No one seemed saddened or angry about it. The audience was just as involved and engaged as the actors.

"What is the meaning of courtyard?" Charu asked. Many explanations and responses were offered by the students leading to a high degree of discussions and verbal interactions between teacher and children. She proceeded to give them the meaning in Hindi. "*Aangan*, an open area which is enclosed in the house." Houses with courtyards are a common occurrence in India and the students understood immediately. The students had lots to say and were definitely given the chance to be heard. Charu was careful to maintain some order and often checked the students who were interrupting her while she was explaining something: "Please do not interrupt me when I am talking."

Another scene was acted out where Mr. Ali is jumped by the ghost. "Who wants to be the ghost?" Of course, dozens of excited hands went up for the coveted roles. One student got the part of the ghost and another boy who was Sikh got to play the part of Mr. Ali. The students were all enthusiastic and very engaged. In the story, the children's mother knew that they were up to something. Charu took this opportunity to inform her second graders that "God has made mummies very smart. When children do something naughty, mummies always know!"

When I left more than an hour later, the lesson was still going on in full swing . . .

Working in Classrooms with 40 Students

From my conversations with various teachers as well as from the responses to the survey questions, I learned that a "large" class-size was a very relative term. Based on the number of students in the classrooms of the 48 teachers in 16 different schools that participated in this study, class sizes in India could vary from 25 to even 60 students in early childhood classrooms. The average class size in The School was 40 at every grade level, and only the Nursery through first grade classrooms had two teachers to a class. Class size had not been an issue that I was exploring or focusing on in the context of this study. I became interested in it as information from the interviews with Charu, Malvika, and Vasudha began to present deeply enlightening facts indicating how the classroom practice of early childhood teachers addressed various issues regarding curriculum and children's development within the context of large classes.

Charu right away acknowledged the challenge of teaching in a large class, and summed up her experience concisely in one sentence: "We have so many children in the class in India—forty. And sometimes it does get troublesome. But still, it's really a rewarding experience—teaching small children." Being a second grade teacher, Charu had an extensive academic syllabus prescribed to be completed during the school year. The formal assessment of students occurred largely in the form of frequent tests and examinations, and there was a strong focus on workbooks and written assignments as homework in order to prepare the students for exams. I asked Charu specifically about her views on the scope for verbal interactions, discussions, and collaborative work in the classroom, and whether there was time and space for her to even consider such an approach with 40 students. Her response was revealing and indicated a level of commitment on her part as the teacher:

> . . . whenever we start a new topic and are doing written work , we always have an interaction for fifteen minutes . . . I say that this is discussion time.

Today we are going to do this topic . . . They've got time to give their ideas. They can ask questions. Ample time for any kind of queries and discussions . . . I'm very strict where work is concerned. But otherwise I give them a lot of time to chat with me. I tell them about myself. I let them tell me about themselves . . . [the relationship] should be more warm and interactive, I think. They should be able to come and ask anything.

Certainly this was very apparent in my observation of her classroom. In addition to making her teaching as interactive as possible, Charu explained that she also took care to ensure what she taught was getting through and making sense to all the students in her classroom:

. . . as a person I'm quite a perfectionist. Probably as a teacher also. I don't want anyone not to learn anything. I want to make sure that everyone knows [learns] everything that I'm trying to teach, you know. I spend more time . . . I go a little slower in my syllabus. But I want to make sure, I know that maybe all forty of them can't learn, but at least thirty-five should know what I am teaching them.

She described how she made her lessons as activity oriented as possible. For instance, when she was teaching about Weather, Charu brought in instruments such as a weather vane, thermometer, and barometer to explain the concepts better to the children. She even took her entire class of 40 students outdoors and used streamers to talk about wind direction. When she was teaching about Foods, she brought in wheat grains and flour and the class made the dough for *chapattis* (Indian bread that is thin, flat, and round) and rolled it out. Charu also described how she tried to make the teaching of English as interesting as possible: "Even things like words—I'm explaining a word, never just the meaning. We first act it out. Then we explain the meaning. Then I tell the children to say as many sentences as they can. So like that it takes at least an hour or an hour and a half doing just five to ten words so that those words are really clear in their minds." The very fact that dramatic play could be used as a teaching tool even in a class of 40 students proved to me that curriculum in large classes need not be overly rigid and structured without any scope for children to participate in activities and verbal interactions. It was not the size of the class but rather the teacher's disposition and attitude that seemed to determine how meaningful and lively the lesson would be.

Throughout the course of my interviews with her, Charu repeatedly emphasized the importance of giving individual attention to the students regardless of the numbers: ". . . you have to give each child time and as much attention as is possible in that situation. And each child needs hearing [needs to be heard] even though you may be rushed for time." From her words, I got the sense that the large class-size definitely did not stop her from incorporating good ideas into her practice although it might certainly have been somewhat of a constraint. It showed that teachers had to work harder and stay more committed to keep learning as enjoyable as possible for the students, but there were endless possibilities for good and effective teaching to be accomplished even in a large class.

Providing individual attention and attempting to get to know each student even in a class of 40 was something that many teachers seemed to be able to do to varying degrees. As I entered Vasudha's Nursery classroom of four-year-olds there was a noisy buzz in this classroom also with sounds of talking, laughing, and eating. The room was square with windows all along two walls, and it was bright and cheerful. There were strings of Christmas tree and stocking cutouts hung across the room, brightly painted, decorated with glitter and sequins. Christmas tree cutouts were also pasted on the window panes. The minute I entered the room all the students chorused as though on cue, "Happy New Year, Ma'am. Good Morning, Ma'am!" They said it loudly, earnestly, proudly, and smilingly! It was break time (scheduled break between classes) and they were each having the morning snack that they had brought from home. There were 40 students sitting at 4 large tables with 10 children at each table. Each child was using a place mat made from a sheet of their own art work that had been laminated. The break lasted for about 30 minutes. As it began to end, the assistant teacher, Alka, called out to the students who had finished their snack and asked them to sit on the floor by her chair. She began to lead them in singing the "Barney" song as a transition between snack time and the next activity . . .

After break time, they had their Language class, and 's' was the letter they would be studying. Alka, the second teacher in this classroom along with Vasudha, stood next to a work sheet pinned onto the bulletin board, pointing to the pictures on it of a snake, a sun, and so on, and going over the sounds of the letter 's'. She then, with Vasudha's help, distributed work sheets to all of the students.

Two bowls of crayons in different colors were placed on each table. The students began their work. All along Vasudha encouraged them to think of other things that begin with the sound of 's.' The students responded in different ways—"sea water," one child said. Vasudha responded, "very good." Another child said "sipper." Once again the teacher responded by saying "very good." The students were talking to each other excitedly, but they were also intently choosing colors and coloring their work sheets. Both teachers walked around the room offering comments and instructions, "Very good. I want to see nice and neat coloring." One little boy said— "Ma'am, look!" and Vasudha affirmed his efforts. There were clear boundaries, with no confusion or chaos. The students seemed comfortable, busy, and happy. Halfway into the class, Vasudha sat down at a table and looked at the work each child was doing at that table. She offered individual feedback, responding to each one, and signing off on each work sheet.

Later that morning when we spoke, I asked Vasudha about one of the boys in her class whom I had observed as being inattentive and distracted. Surprisingly, given that there were 40 children in her classroom, Vasudha knew important details regarding his family and home life. Her explanation as she described the background situation and his home life was moving, and it was evident that she felt a great deal of empathy toward him. She acknowledged that he exhibited aggressive behavior in the classroom but she did not present him as indicating that he needed any special services. Vasudha did not feel the problem lay with the child and viewed the situation as something that could be easily addressed with extra attention, and it was nothing that his own mother could not provide him. There was no reproach or accusation in her words, just an acceptance and recognition of what might help:

> The background is he is an extremely sweet child. Very innocent. He's just completed four years . . . He has extremely good, down to earth parents. But the problem with the child is that the mother is not able to spend a lot of time—the mother is back home only by eight p.m. By that time the child is tired and he goes off to sleep. Saturdays are also working. She gives him maximum time on Sunday. Now he's all the time trying to attract our attention. He's doing something or the other to attract our attention. We've tried our best—but if the mother . . . will be able to devote some time to the child—if she can take a year's leave or something—then I'm sure the child would be able to change. He's an extremely innocent boy—very innocent. When I ask him "Why were you naughty?" he says "I'm a naughty boy, I hit everybody." And he does hit everyone. He's been biting, he's been hitting. He's doing something or the other. He's extremely naughty. And now he's becoming boisterous also . . . and the mother feels guilty—she knows that she is not able to devote time to him.

Vasudha also reflected on her own experiences when she was in school, and emphasized how much her teachers cared for their students and how well they knew each of them, in spite of the large numbers of students in her classrooms. Another example that illustrated the handling of a large class-size lay in Vasudha's description of how each class rehearsed for the Nursery School's annual day celebration. Typically, most schools in New Delhi have an annual celebration that takes the form of some sort of a performance or production in music, dance, sports, and so forth. In this particular instance, all the nursery classes, each with 40 students, were going to participate in a version of a Disney musical. Despite the fact that during my teaching days in India, I had myself experienced the enormous organizational effort that was involved in planning and rehearsing for such events, the idea of more than 200 four-year-olds participating in a musical production seemed a daunting prospect. I asked Vasudha to explain further how this would be handled, and she said, "it's not very difficult. We've got used to it. We take a few children at one time. Get them to practice. Then the next lot comes—the other teacher manages the first lot. Or else if the whole class is sitting together, then they enjoy watching while some children are rehearsing. So it goes on, and it's not difficult. And we give them two months to practice." This description also explained the earlier reports of how the teachers at various grade levels conducted other multicultural activities, holiday celebrations, and all-school assemblies with all of these events encouraging a 100 percent participation for the students.

However, it was not as though the teachers did not feel the frustration of working with so many students. Charu mentioned that in her class there were some boys who were really smart and she wished she could have more time with them to provide a more challenging educational experience for them:

> I've got three-four boys who are really smart. And I wish I could have much more time just with them. Because they come and ask such interesting questions and I want to explain to them and increase their knowledge. But you can't because of our numbers and time constraint . . . In fact they're trying to do things differently from the normal format that we follow. Even when we do written work in question/answers, they want to do it differently. I've never stopped them though I know that that's not the prescribed pattern . . . all of them are so smart.

As I listened to these teachers articulate their classroom practice, it became clear to me that from their perspectives the large number of students in one class, although limiting, was not as big a constraint in the teaching–learning process as might be perceived from a progressive educational perspective in the United States. All 48 teachers who participated in the research reported to have class sizes that ranged from 30 to 56 students per class. The average class size of a private early childhood classroom in New Delhi based on my research data worked out to be 43 students. From all of the descriptions of their classroom practice that they shared with me, I find it reasonable to say that Charu, Malvika, and Vasudha were able to work quite effectively and successfully in their classrooms of 40 students each. I based my conclusions on such indicators whereby:

- the teachers were able to complete a large academic syllabus and work load for the school year as prescribed by the mandated curriculum;

- even within the rigidly prescribed academic curricular schedule the teachers were able to be flexible and create opportunities to make the school experience as activity oriented as possible for the students;
- the teachers appeared to know the individual students and families in their classroom, and recognized that students had different strengths, abilities, and backgrounds. An informal system of differentiated instruction was put into place based on a close collaboration between the teacher and the family. Students who needed more help in their academic learning were usually provided that extra work and support at home through private tutoring.
- in addition to the rigorous academic syllabus, the teachers were also able to implement a nonacademic curriculum with plenty of social and extracurricular activities;
- despite the large numbers, all students were encouraged to participate in school assemblies, holiday celebrations, and other sociocultural events held in the school; and all teachers were required to take turns to coordinate these events;
- despite the pressures to complete the academic goals for each school year, the teachers prioritized the teaching of values and attitudes, which they saw as an integral part of their role, and created the time and space to do so. Besides high academic performance, the teachers viewed it as their responsibility to also promote an overall healthy development of the child's character and personality.

Other indicators of successful teaching in large classes were reflected in the survey questionnaire completed by several additional early childhood teachers working in similar private schools in response to the following question: *At the end of an academic year, do you feel that you have accomplished your goals as a teacher? How do you know?* Some of the responses to this question are summarized here and reflect a variety of criteria by which teachers measured the success and effectiveness of their own teaching:

- By the enhancement in children's attitudes and behavior
- By the smiling faces of the children and their keenness to come to school
- By the fact that children come back to visit with fondness and affection even after they have graduated to the next class
- By the encouraging feedback received from the children's parents
- From compliments received by the class teacher in the subsequent year
- From evidence of improvements in the students' academic progress report
- By their ability to complete the academic syllabus for the year

- By feedback received from their principals and supervisors
- By being able to achieve a class culture of the teacher's liking
- Through review of annual teaching plans and evaluation of how many of the goals had been accomplished.

A few of the participants responded directly that although no teacher could be 100 percent successful, they did consider themselves to be largely successful. Judging from my conversations with teachers and my observations of their classrooms, their teaching was definitely not entirely in the form of lecturing to passive students, and the goal even in this exam-oriented system was not always rote memorization. There was, in fact, a high degree of interactive, discussion-based, and activity-based teaching evident in these large classrooms.

Small Class-Size: A Construct of Privilege and Power

More than any other it was the issue of large class-size that came directly into conflict with my understanding of the ideals of progressive education, which defined early childhood education in many American classrooms. The core ideals of progressive early childhood education as established by educators in the United States have been based on the notions of equality and freedom, which imply an availability of choice, and the freedom and right of the individual to make independent choices. In a typical American progressive early childhood classroom, these concepts are usually manifested as a child-centered, play-based pedagogy, which emphasizes: multiple activity centers; learning experiences that are activity based as play is defined within the context of physical activity; circle time, which allows individual students to express their experiences and voices; and a high degree of importance given to the rights of the individual child. Naturally, a pedagogy based primarily on these core ideals may be implemented successfully only in classrooms that have an abundance of space, learning resources, classroom materials, fewer students, and where teaching is not constrained by a rigidly prescribed curriculum.

As I struggled to make sense of the phenomenon of large class-size, I observed that in India, the case of large numbers of humans in a given group is not an isolated instance found only in the classroom. It is a phenomenon that characterizes all dimensions of Indian society. India is a country with a population of about 1.08 billion people, and large numbers of people are seen in close proximity everywhere whether in homes with joint or extended families; on the crowded streets whether in large cities, small towns, or villages; in the markets, bazaars, and departmental stores; in classrooms on school and college campuses; in the fierce competition for jobs and employment; or in office buildings and other places of work. Crowds are woven into the very fabric of Indian society, and define the basic context of human existence within that society. A room with a few human bodies would be an atypical feature. One of the first things that hits a visitor to India is the crowds and the sheer numbers of people everywhere. Personal space is defined in a very different way in a crowded area, and people get accustomed to having others in and around their personal space all the time. A large class-size is a reflection of this larger phenomenon but one that occurs within a school building.

When working with large numbers, maintaining a semblance of order becomes key for any activity to be productive and constructive. This might explain why classroom teachers in India appeared to be closely focused on behavior management, class control, some degree of silence, and group disciplinary measures such as taking turns in answering, and sharing materials and resources. This also could be an important reason for minimal physical movement and group project work within the classrooms. However, I observed too that although the 40 students in a classroom were physically restricted in terms of space, and were all engaged in working on one activity or lesson at the same time, there were no signs of quiet submission, cowering, fear, or inertia in the room. Rather, in every classroom I noticed high levels of energy in the form of general enthusiasm and eagerness in the rapid raising of hands as students excitedly vied for a turn to answer questions and give comments; jumped to their feet in their eagerness to participate in the conversations with teachers; walked up to the teacher to clarify assignments without any hesitation; chattily discussed the content of the lessons with the teachers; expressed their own opinions and offered their own solutions to situations. The students, their faces bright and their eyes sparkling, were alert and quick to respond; and there was, amidst a sense of humor and frequent laughter, a tangible feeling of comfort, ease, and energy on the part of the teachers as well as the students. As an observer, I saw children who had to be mindful of the overall rules and regulations governing behavior and performance, and yet had many opportunities to create, choose, and control their own social worlds; form or break friendships as they willed; answer the teacher's questions loudly, softly, or not at all; enjoyed the freedom to walk up to the teacher as many times as they felt the need; engaged in conversations with their partners; and were enthusiastic about writing essays even if the topic was chosen by their teacher. During her reading class, even as Charu read out the story from the book in a very teacher-directed manner, there was a look of interest and engagement on the faces of the students. Perhaps, as Viruru (2001) tries to name this experience, it is the rhythm of the whole class participating together in every sentence that brings a sense of shared enjoyment.

The enormous level of verbal and intellectual energy that I observed in these "traditional" classrooms was very different from the more physical energy I have observed in many American progressive early childhood classrooms. For me this raised another critical question of whether one kind of energy was more important than another kind of energy, and if so, who made that decision. How important is the evidence of students' movements and physical energy in a classroom located within a country where physical and outdoor activities have lower priority as compared to a classroom in the United States where a very high value is placed on sports and outdoor activities? And conversely, is it reasonable and logical to expect a high degree of mental energy in a classroom in India where intellectual development is a priority? Are the 40 students in an Indian classroom, which is characterized by low physical energy but high mental energy, any less engaged, or being offered any less of a learning experience, than the 15 students in a classroom in the United States, which is characterized by a higher degree of physical momentum and energy? And if so, who defines good or not so good forms of energies displayed by children, and the correspondingly appropriate or inappropriate learning experiences, learning styles, teaching styles, and modes of engagement? The concept of one-on-one interaction with the teacher and logical reasoning during conversations did not seem to have a significant place in classrooms such as those I observed in India. One-on-one interaction works best when there are a few people in an environment, but one needs a different way of thinking and interacting in order to survive in an environment that is teeming with people and multiple realities, and when the participants in a classroom are called on to handle many different issues and respond to many different needs simultaneously. When viewed in this light would a small class-size be unnatural in the context of the child's life outside the classroom environment, and thus, could it in any way be counterproductive to the learning and growth of the child being prepared to succeed in Indian society?

The class-size issue has been an area of extensive research in most technologically advanced and wealthier nations such as the United States and the United Kingdom Several studies have indeed indicated that small class-sizes can have positive and beneficial effects on student learning. However, despite these findings, empirical data has failed to offer a consistent and integrated explanation for the same. Zurawsky (2003) has suggested that the benefits of a small class-size are conditional and has provided an outline of the specific conditions that must be met in order for small class-sizes to have maximum benefit: (1) early intervention must be provided as soon as the requirement is recognized; (2) the number of students must not exceed 17; (3) if resources are inadequate they must first be made available to at-risk students; and (4) students should experience small classes on a daily basis for at least two to four years. Unless all of these conditions are met, the small class-size will not have any appreciable benefit. It is obvious that this arbitrary number of 17 as the ideal class size has been derived within a given culture where certain specific skills and attitudes are valued over others and where resources are readily available. Viewed from this perspective, it might be impossible to define an ideal class-size that would be universally beneficial. In India, even 40 children may be considered a small group as compared to some Indian schools with 60 children in a classroom. Small class-size in many Indian schools is not currently an option, and early childhood teachers are expected to proceed with their pedagogical practices aimed toward students' high academic performance, successful school experiences, as well as the development of overall healthy personalities despite the constraints of numbers, space, material resources, and the pressures of a prescribed curriculum.

Growing Up in a Large and Competitive Society

Most children in Indian society grow up to live in very crowded and competitive environments in home, educational, or work settings. Charu referred to Indian society as "such a competitive world. Everybody wants to become something. Probably that's why they put so much pressure on their children—there's so much competition. I don't blame parents also. It was so much easier to get admissions earlier—jobs. Now it's so much tougher. Cut off marks [minimum scores for entry into professional colleges] are also so much higher. But still I don't pressurize my children so much. *Theek hai* [okay]—there's a limit. You do your best and that's it." Given that situation, could it be that school experiences in large classrooms as described would actually foster the skills that are required for a life lived in competition and numbers?

Vasudha, in describing to me her personal journey in becoming a teacher, related the determination with which she had to struggle to appear for the entrance examination in order to be admitted into the B.Ed. program and become a teacher:

> After my [college] degree I had to take the entrance exam for B.Ed. On the day of the exam it happened to pour like cats and dogs. There was water

above my waist. But still the exam had to take place. It was scheduled for that day. My father and brother said that they can't even take out the car. But I had to go. So I walked and walked and walked. I was fully drenched. There was no auto [autorikshaw, a three wheeler], no taxi. Nothing could be done . . . finally I somehow managed to get a bus . . . I was about twenty-five minutes late for that exam . . . I sat for the exam . . . but my name didn't come on the first list [did not appear on the first list of successful candidates]. I was very—really unhappy . . . but my name was on the second list . . . So then I did my B.Ed.

Malvika also shared her experience and struggle of facing competition and not being able to take up the career of her choice. It was an enormous setback for her and she described that it was her family's support that helped her through:

I wanted to be a doctor. And I think right from day one, when I understood what a profession is, I wanted to be a doctor. I really tried hard for it, but then I couldn't clear my Pre-medical tests. I cleared my Pre-dental tests, but I never wanted to be a dentist . . . And I remember that for three months I was under a sort of depression . . . I had worked so hard for this thing, and in two years I had not gone anywhere . . . And then my *Tayaji* [father's elder brother] told me that it's alright. Whether you're a doctor or a dietician, it's the same thing [meaning it doesn't matter what career one chooses] . . . Then I became a teacher. And now I'm actually very satisfied with my job. Somehow that question of being a doctor is no more in my mind and it doesn't trouble me any more. And I'm comfortable, wherever I am.

These stories of personal struggle, disappointments, or successes will sound familiar to most who are and have been students in India. They are stories that describe the stress that is typically experienced and endured by most students after completing high school and competing for admission into any professional degree program whether it is in the field of medicine, engineering, law, teaching, accounting, and so forth. The competition for the few available seats for the vast number of applicants is enormous and usually no allowance is made for any delays or illnesses on the part of the candidates.

I come back to asking whether educators who have trained and worked in progressive classrooms in the "West" underestimate the quality of learning experiences offered to students in the average classroom in India because of the large class-sizes. Perhaps the question should be asked whether, in fact, it would be appropriate for children in urban Indian classrooms to learn the skills of not only surviving but excelling in the face of constant competition, which reflects the realities of the lives lived daily by people in urban India. From what I observed in the classrooms, the degree of learning and engagement exhibited by the students seemed to be no less in their large classes of 40 than what I have observed in small progressive classes of 15 in New York City. The students certainly exhibited a high degree of competence in terms of socializing, comprehending the lessons, arguing and debating answers amongst each other, being attentive in order to keep up with the pace of work and competing healthily with their peers for the teachers' attention. Who was to define the appropriateness of the skills that children developed in a large group, or in what way would the skills developed in small classes in order to succeed within small groups work to their advantage in a larger and more competitive playing field?

The concept of small class-size is rooted in learning theories that have been driven by psychological and developmental paradigms in progressive early childhood education in the West, shaped by Euro-American middle-class values (Canella, 1997). This very Eurocentric view of the development of the individual in the context of a small group with a high degree of teacher attention cannot be applied to a large group of Indian children in a different cultural context. Given that the concept of small class-size as being ideal is primarily from the perspective of a rich and privileged society, another dilemma that emerges is whether a small class-size could even be an option in the Indian context where resources are scarce and people are many. The issue of small class-size and the quality of one-on-one interaction between the teachers and the students becomes further irrelevant when the real issue in many Indian classrooms is a struggle to acquire the foundations of literacy amidst a paucity of resources such as space, pencils, paper, books, crayons, and even food and clothes. Changing the class profile to one with a smaller group of children can realistically be done only in private and financially well-off schools that have the resources for more physical space as well as for buying more materials and hiring more teachers. This would only result in widening the gap between private and government schools thus highlighting the socioeconomic divide even more. One would have to carefully consider the benefits of reducing class sizes only in the wealthier schools compared to the disadvantage of a corresponding reduction in the overall egalitarian objectives. Large class-sizes probably will continue to be the norm in Indian schools and the situation can change only after the real challenge of population control is addressed.

Indian children are not raised in sparsely populated towns, homes, families, and communities. Their growth happens in a context that has a high degree of social interaction dominated by intergenerational experiences. Children learn to wait for their turns, and to share their meager possessions of toys and books with siblings, cousins, and neighbors; they learn to value the scarcity of commodities such as picture books, art and craft materials, and so forth. Before educators argue to apply the small class-size standard to the Indian classroom using a Euro-American educational discourse, it would greatly help the general understanding if they were to consider (1) what the educational aims for a populous Indian society are and how they might be different from those of wealthier, culturally different societies with much smaller populations; (2) whether those educational aims have been shaped by Western educational philosophies or are more in consonance with the local values of the Indian culture; and (3) what are the specific skills and attitudes that will allow children in India to succeed and live competently in the large, urban, and densely populated sections of the nation.

In conclusion, I do acknowledge that although the teachers in my study appeared to be quite effective in their classroom practice, teaching is not necessarily successful or effective in all classrooms with large numbers of students. But then teaching is also not successful and effective in all classrooms with small numbers of students as research has shown. My observations and experiences lead me to believe that effective teaching depends to a large extent on the dispositions of teachers, school administrators, families, and children and their attitudes toward education in general. I noticed that the attitudes and dispositions of the participants in my study were defined by a common cultural ethos that enabled them to strive toward a common goal:

- there was a general acceptance that large class-size was a non-negotiable norm
- there was a positive attitude and a strong commitment on the part of the teachers to making it work
- many teachers viewed their students as their own children and were deeply invested in helping them develop into strong, healthy, and competent human beings
- most teachers held their students to high expectations and wanted them to do their best in everything
- teaching, in general, was considered to be a noble profession and there was a high value attached to the concept of education
- teachers, in general, were respected and valued by students and their parents as well as their extended families
- teachers as well as parents/families of students wanted students to learn good values and to perform at high academic standards. This resulted in children being "taught" and held to the same expectations at school and at home.

Research has shown that the biggest advantage of small class-size is greater individualization (Molnar et al., 1999). From what I learned and already know, private tutoring is commonplace in Indian society and many children are tutored in their academic work at home. Private tutoring can take many forms from expensive services provided by institutionalized bureaus of tutors to free tutoring provided by friends and relatives of the family, and even by older siblings of the student in many cases. I perceive this private tutoring at home to be the equivalent of focused and individualized teaching that the student benefits from. Research has also shown that small class-size does not automatically improve learning and teaching behaviors, and in fact can lead to more interrupted teacher–student interaction as children expect to have their demands met and questions answered immediately. Teachers have to work just as hard in small classes to manage learning effectively (Blatchford et al., 2002). It seems that the early childhood teachers whom I observed and interviewed had somehow taught themselves how best to manage teaching and learning in their large classes and make it as effective as possible given the tools with which and the circumstances in which they worked.

Developmentally appropriate practices and the related notion of the emergent curriculum are not concepts that can be readily practiced with a group of 43 students. It would be nearly impossible to implement a curriculum that could be driven by the individual stages of development in all the developmental domains of 43 children or the interests of 43 children. Further, as Kumar (1992) suggests, children's interests are constantly changing, and to keep track of changing themes for a group of 43 in order to design a curriculum would be a formidable task for any teacher. This makes it necessary for the use of a structured and preplanned curriculum, which is, in fact, the norm in all schools in India. However, I do believe that what is more important than the particular type of curriculum used is the consistency between school and home and the value each has for education that ultimately results in successful school experiences. During my interviews with the teachers as well as the school administrators, I sensed an underlying assumption that teachers and parents expected the same educational objectives for the children in terms of values and right attitudes, intellectual development, and academic proficiency. Due to the widely existing phenomenon of the joint and extended family systems, most children's social and emotional growth takes place largely within their home environments, and schools are thus able to focus more closely on the academic curriculum. Chapter 9 presents a more detailed discussion of the early childhood curriculum.

Chapter 9

A Socioculturally Constructed Early Childhood Postcolonial Curriculum: The Interfacing of Three Culturally Different Educational Discourses

A system of education is usually derived after answering questions such as: What is the aim of education? What are the activities that will achieve that aim? What are the learning theories that will govern the activities? What are the teaching and assessment strategies that will be used? What is the nature of the roles of the students, teachers, and administrators? The answers to all of these questions form the educational philosophy of a school or society. Educational philosophy in the West has been variously conceptualized such as a general theory of education by John Dewey or as a screen for selecting educational objectives by Ralph Tyler (Wiles and Bondi, 1989), or the application of philosophical ideas to education (Ozmon & Craver, 1995). In attempting to identify an educational philosophy it is necessary to confront value-laden questions regarding the underlying values and beliefs that a community upholds. These questions arise from greater philosophical issues about the nature and purpose of life, the roles and abilities of human beings, and worldviews that express what is good, what is true, and what is real. Ultimately, it is this overarching philosophy that is reflected in any learning or psychological theory or school curriculum, whether Indian or Western.

Throughout this book I have maintained a discussion on Indian philosophy and psychology as well as Euro/American learning theories and psychologies. This at first appears to be a comparison between philosophies on one hand, and learning theories on the other. But when an attempt is made to identify the underlying philosophies that shape Euro-American developmental/learning theories, a situation of apples being compared to oranges can be avoided. Wiles and Bondi (1989) identify five basic kinds of educational philosophies within which the most common Western psychological theories of development and learning can be grouped. I have briefly described the five philosophies to illustrate how a theory of learning or development can emerge out of a particular school of philosophy.

The first philosophy is Perennialism and this is based on the idea that the distinguishing characteristic of human beings is their ability to reason, and that the purpose of education should be the development of rationality through the teaching of eternal truths. The teacher's role is to interpret and tell, and subjects are taught through disciplined drill.

Another philosophy is Idealism which holds that ideas are the only true reality, and reflects the refined wisdom of men and women. The aim of education is to search for true ideas and character development. To achieve this the students must be able to sharpen their intellectual processes by learning about ideal behavior, which is to be modeled by their teachers.

The philosophy of Realism rejects the notion that ideas are the only reality, and acknowledges, rather, that reality exists in the world as it is and that education should teach students the laws of nature and the physical world by developing their skills of observation. The curricular content influenced by this educational philosophy would prioritize facts and knowledge about math and science, and classrooms would reflect order and discipline, like nature.

A fourth philosophy is Experimentalism, also known as Pragmatism, and this takes into account that the world is in a constant state of flux and change, and reality lies in actual experiences or interactions with physical and social worlds. Such an education would prioritize social studies and physical experiences. Hands-on, practical, inquiry-based problemsolving would be the basis of the curriculum and the teachers would be facilitators of the students as learners.

Existentialism as a fifth philosophy recognizes that goodness, truth, and reality are individually defined and driven by internal impulses. Schools based on this belief would allow students to discover themselves and their place in society, and the teaching would focus very strongly on individualization and the individual interpretation of arts, ethics, and philosophy.

Ozmon and Craver (1995) discuss Eastern Philosophy as an additional philosophical paradigm. This includes Indian, Chinese, Japanese, and Middle Eastern schools of thought that have influenced the educational philosophy in those respective regions of the world. Most of the European philosophies had their origins in the ideas of Ancient Greece, by which time philosophy had already reached a high level of development in both India and China. While Western philosophy emphasized science and logic, the senses, and materialism, Eastern philosophy emphasized the individual's inner world and intuition (Ozmon & Craver, 1995); unlike Greek philosophy, which separated philosophy and religion, within the Indian and Chinese schools of thought, philosophy and religion were very closely and almost inseparably intertwined. In its practical applications, Western educational philosophy is seen to emphasize material goods, social advancement, and changing standards, whereas Eastern educational philosophy places a stronger emphasis on the sense of duty, familial ties, and veneration of elders and ancestors. Several educational philosophies in the West also emerged from the philosophical discourses of Heidegger, Nietzsche, Schopenhauer, and Husserl; many of these philosophers had been impacted deeply by the rise of Indology in the eighteenth century (Dhillon, 1997). The Western world's interest in the *Veda* began with the translation of the *Bhagvad Gita* by Charles Wilkins in 1784, and it is relevant to keep in sight the fact that the interaction between Indian ideas and Western philosophies was established several hundred years ago.

Postcolonial Interpretations

In my discussions on postcolonial perspectives in chapter 1, I made reference to the term "postcolonialism" as a research paradigm that seeks to situate contemporary educational issues in the context of underlying colonial experiences. From this vantage point, the colonized condition may be described variously as: a powerful interdependence between the colonized and the colonizer (Gandhi, 1998); a continuing contest between the dominance of the colonizers and the consequent legacies that were created (de Alva, 1995); the intercultural negotiation between the ideas of the colonizer and the colonized (Pratt, 1992); a transaction, a two-way dialogue between the philosophies of the colonized and the colonizer (Trivedi, 1993). This conceptualization of the hybridization of and the transactions between ideas, discourses, and pedagogies was evident in the interplay between the teacher's preparation and their practice, and between their learning to become teachers and their actual work in the classrooms. In this chapter I describe how the different pedagogies that I observed might have originated and evolved, and how the school's childhood curriculum reflected these various influences and discourses.

Education as a systemic phenomenon in India appeared to be complex, multilayered, and confusing and the early childhood curriculum that

I observed could not be named as it did not fit into any one identifiable model or educational philosophy. Judging from the data, a unique kind of early childhood curriculum was being enacted: the goal of teaching academics and developing academic proficiency in young children worked hand-in-hand with the goal of teaching values and developing good character, and at the same time there was also the intention for some child-centered practices as defined by the tradition of progressive education. Further, this curriculum was being implemented within a school environment that had large class-sizes and a high level of energy and engagement on the part of the students as previously described. The concept of cultural hybridization (Bhabha, 1994) of diverse elements—ancient and modern; traditional, colonial, and progressive—was very apparent in this socially and culturally constructed curriculum. After trying to tease apart this curriculum, the closely intertwined strands from three distinct discourses could be identified in the classroom curriculum of all three early childhood classrooms included in this study. These discourses represented: (1) the deeper underlying educational purposes that represented ideas from Indian philosophy; (2) the policies that had been put in place by British colonial administrators in colonized India; and (3) traces of some core concepts from progressive education that are commonly seen in many American early childhood classrooms. These findings raised further questions as I began to look for the relationships among the *how*, the *what*, and the *why* of this early childhood curriculum (New, 1999). How had the present curriculum in these early childhood classrooms evolved? What specific early childhood educational objectives did teachers work toward by way of this curriculum? Why did practitioners choose to select these particular educational objectives? In what ways did this curriculum differ from one that is held to be "appropriate" as per the definitions provided by the dominant Western early childhood discourse? And how were the boundaries of philosophies and educational theories shaped by culture, history, and politics navigated in the practical implementation of the classroom curriculum? Some of the answers were found in the voices of Charu, Malvika, and Vasudha as they described aspects of their practice with a multifaceted curriculum.

As described in chapter 4, the data had revealed interesting responses from the participants in this study with regard to the four aims of education they considered to be most important: (1) the teaching of values and "right" attitudes; (2) the development of the intellect; (3) fostering cultural diversity; and (4) teaching academic proficiency. Based on the discussions and analysis of research and historical literature presented in previous chapters, it seems that three of the four aims of education, namely the teaching of values, fostering intellectual development, and fostering diversity, could be traced back to Indian philosophical beliefs. Rigorous academic teaching and a strong focus on textbooks, tests and examinations, and basic skills such as writing, reading, and math could be traced back to the influence of the colonial policies in India. In other words, the curriculum appeared to be multidimensional drawing from multiple discourses. One dimension of this curriculum was highly structured, content-based, and had a formal academic aspect that was prescribed and mandated by the Indian government at the national level, and which had its roots in the colonizing influences of the British government on Indian education. Another dimension was the ongoing, parallel values-based curriculum inspired by the teachers' tacit knowledge and implicit beliefs resulting from their sociocultural-historical constructivist learning influenced by an all-pervasive Indian philosophy, and supported by the school administrators (Gupta, 2003). A third dimension, although not specifically mentioned in the articulation of educational objectives, was related to the urging by school administrators, and to the teachers' making sense of, the ideals of progressive early childhood pedagogy especially in schools that were financially well-off and had been exposed to Western ideas in education by way of workshops and international teacher exchange programs.

I discuss the social–cultural–historical nature of this curriculum by exploring each of the three dimensions chronologically, starting first with the influence of the discourse of Ancient Indian philosophy, proceeding to the discourse of British colonization, and lastly touching on the discourse of American progressivism, demonstrating how each was manifest in these early childhood classrooms.

Influence of Indian Philosophy on the Early Childhood Curriculum

The ethos of Indian philosophy has been analyzed earlier in the book in terms of a general way of life, and more specifically seen in the teaching of values and attitudes, developing the intellect, fostering diversity in classroom situations. There was the definite influence of Indian philosophy on the teaching of values. Chapter 4 presents in detail the values that were specified by the practitioners. For the purpose of this discussion some examples of the values mentioned by the teachers were: (1) respecting one's elders; (2) being mindful of one's duties and responsibilities toward society and family; (3) practicing virtues such as kindness, sharing, caring, helping, and so forth; (4) being able to relate well with others at a social level; (5) being able to adjust well and adapt to changes in life; (6) and being concerned for the environment and its protection. Although many of these values are not very different from those typically stated as educational goals by teachers in the United States, a direct connection may be seen between these professed values and the ancient "lived" philosophy of Hinduism, which, as I mentioned earlier, offers specific recommendations for living a good life. Chapter 4 includes a detailed description of the connections I have made between the aims of education as articulated by the participants in the study and the all-pervasive ideas of *dharma* and *karma*, which are central to the thinking of Hindus as well as many non-Hindu Indians. Influenced largely by their upbringing, background, and a system of prior beliefs and practical knowledge and not as much by their professional training, teachers and school principals identify basic concepts within this worldview as important educational aims for young children.

The influence of Indian philosophy was also seen on the educational aim of developing the intellect of the child. Intellectual development and sharpening of the intellect has its roots in Ancient Indian philosophy too, as described earlier. This perspective seems to imply that it is the intellect that gives reign to the individual's social and emotional development. Shakti's views, along with Charu's emphasis on helping children conduct themselves with emotional balance and social awareness of their duties toward others, are consistent with the Indian understanding of moral development, which is the primary focus of two of India's greatest writings: the *Bhagwad Gita*, which describes Arjuna's moral dilemma; and the epic *Ramayana*, which is the story of Ram's voluntary exile in accordance to a vow that his father was forced to make. The epics relate how both men adhere to their *dharma* or duty. Activities related to both the *Ramayana* and the *Gayatri Mantra* were seen to be a prominent part of the curriculum. Vasudha, a Nursery (Pre-K) teacher participant described how she, when it was her turn to conduct the Nursery school assembly, chose the epic *Ramayana* as her topic:

> . . . I had *Ramayana* Assembly this time and I had written the whole script on my own. And I'm not much into Sanskrit or anything. But then I went down to the children's level and I made those *chaupais* [four-line verses resembling the actual format in which the story of *Ramayana* is written] according to the children—so that they could understand . . . I had written the lyrics for their age. I was so happy that I had written something on my own and something that the children could relate to and have really enjoyed . . .

This example is also a good illustration of how the teacher establishes "learning goals and expectations for children that correspond to the sociocultural tools and practices valued by the larger society" (New, 1999, p. 268).

The third influence of Indian philosophy was seen with regard to the teaching of respect for cultural and religious diversity, yet another important educational aim for the teachers. Over the course of her more than 5,000 years of history, India has been known to be a nation of vast diversity and has had a long tradition of nurturing populations from different religious groups. People in India belong to different racial and religious backgrounds and the Indian culture has continuously been enriched due to the ready assimilation of diverse cultural elements. In India, religious holidays such as Christmas and Easter (Christianity and Catholicism), Eid (Islam), Diwali and Holi (Hinduism), Buddha Jayanti (Buddhism), Guru Nanak's Birthday (Sikhism), Mahavir Jayanti (Jainism), to name a few examples, are holidays on the national calendar and many schools are closed on these days.

Efforts to promote diversity are evident in many aspects of school life in urban India. A common example of the acceptance of diverse perspectives is seen in the manner in which private schools print their annual school almanacs (calendars and homework diaries that are used by students on a daily basis through an academic year) with quotations and prayers from different religions and philosophies to inspire students. Charu described how she had incorporated the celebration of diverse holidays in her second grade classroom despite the constraints of her academically prescribed curriculum:

> For *Diwali* . . . we have a *diya* [oil lamp made of clay] making competition, and *diya* decorating and *Diwali* cards, and we light *diyas* in class. For Christmas, as you saw, we had been decorating . . . the class and all. So we do try and incorporate most of the festivals . . . we have two Sikh boys . . . I made sure that when it was *Guru Nanak's Birthday* [a major Sikh holiday in India] they all knew what it was, and after that I told those two boys to find out whatever they could and tell us about *Guru Nanak*. So they [researched the holiday and] read it out . . . and how they celebrate it also they talked about. And I have a Muslim girl also, who did the same when *Eid* came . . .

In addition to this approach in the classroom, and as I have described in an earlier chapter, a three-language formula as part of the national school curriculum in India ensures that the majority of schoolgoing children in the country are at least bilingual, with many being trilingual. The three languages that students are required to learn during their school careers are the regional language of the state in which they reside, English, the official language of India, and Hindi, the national language of India. Thus, under constitutional law, a form of multiculturalism and multilingualism is incorporated in the school curriculum, and becomes translated into a form of pedagogy that observes religious holidays from different faiths and implements cultural activities from diverse ethnic groups.

It is quite apparent how the three aims discussed above are intimately connected to the overarching Indian philosophy. Not only are the concepts of *dharma* and *karma* seen to be central to these objectives, but they appear within the Indian worldview as being fundamental to the domains of child development such as social development, emotional development, and moral development. This approach, which considered a child's growth and development as a complex, three-dimensional whole without different theories governing individual developmental domains, seemed to find support in the findings of the study that indicated that developing an overall, balanced, all-rounded social, emotional, and intellectual personality in the children was an important objective for the three teachers. From a theoretical standpoint, this can also be supported by the sociocultural emphasis of the Vygotskian approach. Unlike Piaget and others who saw human development as proceeding in maturational stages, Vygotsky viewed development as proceeding through a social process without stages, where learning was mediated through interactions and the social phenomenon of language (Williams, 1999), influencing children's construction of meaning and subsequent social attitudes and behaviors.

The Influence of the Legacies of British Colonialism

This influence could be seen in the larger system of education prevalent in India and is directly related to the educational aim of teaching academic proficiency. In India, teaching to the test becomes prioritized due to the exam-oriented nature of the current system of education. As previously described, India came under the rule of a series of foreign powers over time such as the Mughals from approximately A.D. 1100 to A.D. 1600, and thereafter the British until 1947. Under these influences, several systems of education were implemented and elements of these can still be found in the multifaceted educational system in India. The early seventeenth century saw the emergence of missionary schools in India with their predominantly Roman Catholic beliefs and Christianity's focus on individual salvation. Later, as the British gained supremacy and control over all of India as a colony, educational approaches in the scientific methods of modern Europe were implemented.

Historically, two educational policies implemented in India under the British rule facilitated the establishment of the "West" not only in structures but also in the minds of the natives: (1) under the British administrators a system of education was designed and implemented that would bureaucratically control the way native Indians would be educated, and (2) the then viceroy of India, Lord Macaulay, proclaimed that the English language would be made the medium of instruction in schools in India. These two policies led to the system of education in India becoming strictly examination oriented and textbook driven, where both the tests and the textbooks were designed by the bureaucrats. This in itself would become a lasting legacy of the British colonial era. The creation of a class of native elites in its own image was another legacy of British colonialism as Lord Macaulay pressed the urgency of forming this middle class who would serve as interpreters between the British rulers and the millions that they ruled (Varma, 1999). India's current educational system continues to be driven by a strong focus on the assessment of students through a rigid examination system. A mandated national academic curriculum is translated into a prescribed syllabus and corresponding textbooks for each grade level in schools. Students are individually evaluated at the end of each academic year for the content of the textbooks they have studied. As each year's academic content is built upon the previous year's syllabus, the pressure to develop proficiency in academic skills filters down into the early childhood classrooms also. Even in the lower grades the curriculum is organized into content areas such as Math, English, Hindi, and Environmental Studies.

The outline of a typical academic syllabus for private school nursery classes for four-year-olds in New Delhi might look like this:

English: Recognition of capital and small letters, A to Z
Phonetic sounds of letters A to Z
Using phonics to sound out simple words
Recognition of sight words
Pre-writing skills
Pattern writing and free hand drawing
Writing of capital letters A to Z
Math: Recognition of numbers 1 to 9
Value of numbers 1 to 9
Oral counting from 1 to 50
Writing numbers from 1 to 9
Pre-math concepts such as same/different, big/ small, more/ less, before/after, many/few, left/right, over/under, first/last, and so forth

In addition to these academic topics, the prescribed curriculum may also include the exploration of topics such as: Me and My Family; Food; The Five Senses; Personal Hygiene; Plants; Animals; Cloth; Professions; Transportation; Places; and so forth. Workbooks based on these topics are often used in most schools even at the nursery level (Gupta & Raman, 1994).

At the first grade level, a typical academic syllabus would prescribe content areas such as English, Math, Hindi, and Environmental Studies for each month in the school year. The following would be an example of the academic curriculum for the months of March and April, which mark the start of the school year in Delhi:

English: Introduction of the concept of "naming words"
          To understand and practice the structures "is" and "am"
          Creative writing
          Corresponding exercises in the English workbook
Hindi: Study and practice of the Hindi alphabet, specifically of the first vowel sounds
          Oral counting of numbers in Hindi from 1 to 10
          Naming the different parts of the body in Hindi
          Corresponding exercises from the Hindi workbook
Math: Revision of math concepts done in the previous year
          Learning and using math vocabulary
          Classifying according to shapes and sizes
          Number values and writing number names upto 19
          Dodging numbers
          Numbers before, after, in-between
          Making of zero and one
          Corresponding exercises from the math workbook
EVS: The study of parts of the body, functions of the sense organs
          Importance of healthy food
          Corresponding exercises from the EVS text/workbook

The syllabus may also list activities and field trips related to each of the content areas. One result of an academically rigorous curriculum that is examination oriented is that many textbooks have to be used and a great amount of homework had to be done by the students each day. Malvika, the first grade teacher, expressed her disagreement at the rigid and extensive academic curriculum the teachers were expected to complete:

> . . . they [the policy makers] have got a formal curriculum which we have to follow, and we do it . . . at times being in the class I feel that the academic pressure is too much for a Class One [first grade] child . . . We've got too much to be covered in that one month or whatever time the teachers have that we have got hardly any time left to talk to them [the children]. And this is not right . . . Too much testing . . . This is not right. A little bit of testing needs to be removed from our system . . .

Some degree of memorization has always existed in India because of a strong oral tradition. The learning of scriptures, literature, mythology, science, and history was achieved through memorization. During ancient and medieval times, it was critical within communities for each generation to transmit knowledge to the next and thus preserve the "cultural capital."

In modern India, memorization assumed importance in both government and private schools due to the requirement for all students to pass an examination in order to move up to the next grade level. When the curriculum is so examination oriented and textbook driven, memorization becomes an important skill to develop in order to be successful in schools in India. But there was a definite distinction between academic memorization and intellectual development, and both were seen to exist in the school curriculum.

The teaching of values and development of the intellect were not as overt a part of the school's prescribed curriculum as was the teaching of academic proficiency. Nevertheless, the former two goals surpassed in importance the academic goals to such an extent that the principals of several private schools gave important consideration to the personalities, dispositions, and attitudes of candidates when selecting teachers for their early childhood classrooms. Thus, in the classrooms, the ethos of Indian philosophy emerged as a "hidden curriculum" more powerful than the prescribed academic curriculum.

The third influence on the early childhood curriculum that intersected with the influences of Indian philosophy and British colonialism was that of American progressivism.

The Influence of the Ideals of American Progressive Education

Although progressive education might be the most enduring educational reform movement in the United States (Semel & Sadovnik, 1999), the term itself is widely used and can encompass a broad range of ideas and reforms thus lacking a precise definition (Cremin, 1988; Kliebard, 1995). Most educational historians in the United States, however, agree that progressive education can be characterized by a core of ideas such as a belief in individualized, democratic active learning, and child-centered education that is aimed at the "whole child," and a curriculum that goes beyond being bookish and being based on rote learning to meeting children's social, emotional, psychological, and biological needs (Davies, 2002). In many American early childhood classrooms, these core ideas take the form of such constructs and techniques as circle time, multiple learning centers, small class-sizes, individualized and differentiated instruction, whole language approach, thematic units based on children's interests, experiential and project-based learning, authentic assessment, block building, strong social studies focus on neighborhood and community, and so forth.

Although these ideals of progressive education were not specifically mentioned by the participants in their articulation of what they considered to be the important aims of education, the language of progressive early childhood education seemed to be somewhat sprinkled into the conversations of the three early childhood teachers Charu, Malvika, and Vasudha. Some examples of the terms mentioned were circle time, activity centers, desire for small class-sizes, social-emotional development, and the notion of multiple intelligences. Under the influence of modern Western pedagogical approaches accessed through the media, Internet, international teacher exchange programs, and professional development workshops, administrators in private schools are able to urge some control over the early childhood curriculum in their own schools and prevent it from being totally dominated by academic teaching and rote learning. The school administrators seemed to want to expose their teachers to the early childhood techniques of Euro-American classrooms by offering these various workshops. Thus the influence of progressive education was somewhat apparent in the manner in which the three teacher participants described their activity-oriented approach for the teaching of academic content and concepts. However, this may not be the case in many other private schools or in most government schools in India that are less exposed to the Euro-American discourse of early childhood education, and where rote learning of basic reading and writing skills is still the dominant practice. Charu, Vasudha, and Malvika, with 40 children each in their classrooms, recognized the difficulty of developing an ideal child-centered curriculum with large classes. Nevertheless, they attempted to use as many activities as possible in teaching concepts to their students. As all three teachers articulated their classroom practice, many of their descriptions reflected some aspects of the core ideals of progressive education as presented below.

Balancing Academic Work with Activity-Based Experiences

Malvika, the first grade teacher, explained how class activities were scheduled during the school day for the six-year-olds to emphasize that along with academic instruction students also had opportunities to experience learning through activities such as music, movement, and art:

> We work for around five hours, from 7:30 a.m to 12:30 p.m . . . In that let's say we have a twenty minute break time. We have two teachers in one class. We're only teaching [formal academic teaching] half of the time so let's say only fifty percent of the time is given to academics. And then we take them for nature walks, we take them for computers, they go to the activity room

which is a play room where they've got music, dance, art and craft. And while we are teaching, there are two teachers in the class. One is there to teach and the other one is to supervise. Because handling a class of forty to forty-five is not that easy, and with especially such a young group . . .

Charu, the second grade teacher, described the use of thematic units alongside the academic syllabus, and explained how she taught the unit on food using an activity-based approach, but at the same time acknowledging that this approach was constrained by the large class-size:

Like when we did Food, we brought in wheat grains and then flour, and the entire process we did in class. Making *Chapattis* [a type of Indian bread]. We didn't actually cook it. But even then I want to make them learn naturally. Not just knowing what it is. Actually seeing and doing it. So they should actually experience as much as possible . . . Not so much is possible because we have forty children and it really hampers any activity because so much of noise is generated. Plus because we have so little time and so much syllabus [academic] to complete . . .

But Charu gave another example of this interfacing among thematic units, the prescribed academic syllabus, and activity-based teaching:

Like we did Air, and I started by asking the children, "can I just switch on the fans?" So we switched on the fans. And they said, "Ma'am please switch it off. It's so cold!" And I said "but why are you feeling cold? What is it that's happening? We can feel the air. Okay. So let's look out—why are the trees moving?" You know, so for 10–15 minutes I always do it this way. Then I asked them to bring balloons, and they kept asking Ma'am, why? I said, "it's a secret, you'll find out later on." So then I said, "now everybody take out your balloons and blow into it. So what is it that you're putting inside that? Okay—leave it, let it come out, feel it." So for some time I always let them . . . try and experience what I'm trying to teach them.

Vasudha, too, had an example that illustrated the activity-based approach to teaching in her Nursery classroom:

Like this month's topic is Plants and we'll take them out in the garden and have a nature walk, they'll observe the plants, grow their own seeds. They'll sow their seeds and they'll look after their plants . . . But when they come up [back to the classroom] they'll be a follow up activity. Like today we did parts of a plant. They went and observed a plant. They took out the plant, they observed the root and all. And then we did [reviewed] all the parts of the plant [points to a piece of art work that is pinned on the bulletin board]. This is the pasting activity related to the topic of the month. But I don't think this should be included in academics.

Connecting the Curriculum to Students' Real-Life Experiences

Charu explained her efforts to connect the curriculum to the children's real-life experiences but with the awareness that ultimately the lesson had to revert to basic writing skills.

> Like, yesterday was a foggy day. So in the morning we kept discussing the fog and all. What was everything they perceived about a fog? Okay—so they couldn't see the buses from their stops [bus stops]; did they see any accidents; and how should we keep ourselves warm on such days; and what they'd like to eat, and we kept discussing it for almost an hour. Then I gave them their English copies [English notebooks] and I said now you are going to write about it . . .

Individualizing Teaching

Vasudha, the Nursery teacher who taught four-year-olds, spent some time talking to me about the school's efforts in implementing at what we might call differentiated instruction. The challenge was how individualized attention could be provided in a class of 40 children:

> You know there are some children who are very slow in learning English and some who are very good at expressing themselves, and there are some children who are very shy . . . separate attention is given to the children who are shy. There are some parents who are not English-speaking. We try and reinforce the language so that they also speak in English. And we have more advanced teaching aids for the children who are not able to express themselves, or the ones who take a little more time to learn . . . we even keep in mind that there are some parents who are working parents and they are not able to devote a lot of time. We have a puppet making class [where children make puppets with the help of their parents] . . . So for the parents who are working and can't spend a long time, we give simple puppets to them . . . And the ones who are not working, they are given the elaborate costumes . . .

Circle Time in the Nursery Classroom

Vasudha also explained that she implemented circle time in the daily routine of her classroom, allowing the children to share their thoughts and experiences:

> In the morning when the child arrives, the first period is basically warming up exercises. We have a prayer . . . one prayer in English, one prayer in

Hindi. And then after that we have circle time. Circle time means we have any one topic of discussion with them. We can have any topic. Like today the children came back [for the first time after winter break] and [we discussed] what they did during the holidays . . . if we don't have circle time then we have an activity—like pottery and taekwondo [a form of martial arts], and in summer we have water play for them . . .

I was struck by two things here. One, that circle time could sometimes be substituted by another activity implying that the time available for nonacademic work was limited, and second, that the students' discussion was restricted to one topic each day.

Teachers as Professionals

Vasudha went on to explain the teachers' role in determining some aspects of the early childhood curriculum. There seemed to be some evaluation and discussion as the teachers and school administrators negotiated the implementation of new ideas, indicating that teachers had some degree of freedom and decisionmaking regarding the early childhood curriculum:

We sit together . . . [our] view points are taken on whether this [new approach] can be done in our school, will it be convenient for the teachers to handle with so many children? So we give our own viewpoints . . . And for anything new we want to do or we want to implement something, we do have a word. And if supposing the curriculum is not suitable to my class and if I feel my children are not able to do it, I'm free to change it according to my standard . . . I mean, we are encouraged at all times . . .

There was an underlying sense of how some ideas of American progressive education are held to be appropriate and desirable by these school administrators and teachers, and the above examples provided illustrations of the attempts on the part of the teachers to incorporate those ideas into their own practice to varying degrees, alongside the other demands of teaching academics and their own conviction of teaching values to their students.

Postcolonial Hybridity

British colonization in India stretched over a period of almost 300 years, resulting in an intimate juxtaposition of Indian and Western ideas and worldviews, and a continuing transaction between the two. Just as the daily lifestyles of people in India are based upon the all-pervasive phenomenon of coexistence, and reflect elements that are clearly Indian and Western, so also teachers' practice and school curricula were seen to draw upon the multiple coexisting realities of Indian philosophy and Euro-American pedagogies, allowing teachers to ground their classroom practice in their realities and their own practical knowledge. It was clearly apparent from their interviews as well as classroom observations that the three teachers were interweaving elements of an ancient cultural philosophy that was still actively practiced, with elements of British colonial and American progressive educational perspectives. Early childhood education as seen here included the teaching of values and correct attitudes, developing the intellect, fostering a sense of justice and equality through diversity, and the start of academic skills learning. All teachings were interwoven into the overall curriculum rather than kept in separate compartments. Within one lesson the teacher could teach academics, respect for elders, the importance of conserving electricity, and give examples of stories from Indian mythology. Shakti, the school principal, summarized this concept of hybridization when she said that teaching children "is to be done not necessarily through the modes we use like the books and the pencil and the chalk. This can be done by play, by discovery, by experimentations, by—just being oneself. There is so much inside each individual which the Indian philosophy emphasizes."

For the purpose of analysis, I have categorized these elements into distinct philosophical influences, but in the "enacted curriculum" the elements were integrated into one socially constructed curriculum. The discourse of Indian Philosophy and culture influenced both the content of curriculum and the methodology of teaching, and could be seen as the "hidden curriculum" in the informal approaches to teaching and learning in the classroom, and in the more deliberately implemented values curriculum. Specific elements from this discourse included: the teaching of values central to Indian philosophy; the importance given to the concepts of duty, respect, and the observation of customs, traditions, religious holidays; prioritizing intellectual development; the occurrence of multilingualism; the personal philosophies of teachers, children, and their families; the implicit beliefs of teachers; and the image of the teacher and child. The discourse on British colonialism was dominant in its influence on content and methodology, and was manifested in the rigidly structured tests and workbook curriculum. Specific elements included: the nationally prescribed curriculum; an exam system driven by prescribed texts and work books; a system conducive to rote memorization; teachers who were regarded as technical workers; limited evidence of opportunities for creativity for both teachers and children; and the adoption of English, the language of the colonial rulers, as the language of instruction. And finally, the discourse of American progressive education that influenced, although in a limited way, the early childhood teaching methodology in the classrooms observed. It was seen in teachers' attempts in creating an activity room with learning centers that would allow children "choice time"; incorporating "circle time" into the daily schedule; in their desire to work toward "child-centered" practices based on the needs and interests of the children; providing opportunities in the classroom for children to voice their thoughts and experiences; providing abundant classroom materials; and the desire (wishful!) for small class-sizes as per the recommendations of early educators in the United States. However, it might be said that these influences would be restricted mostly to financially well-off private schools in metropolitan cities of India.

During my research and review of documents, I came across examples of texts and readers prescribed for schools and I found their contents quite fascinating. In India, it is usual to find textbooks published as a complete series across grade levels in all content areas such as English, Math, Hindi, Science, and so forth. There are numerous textbook series that are published within the guidelines prescribed by the NCERT and by the CBSE, and schools can then select the series they want to adopt. As a random example of the contents of one such text, I want to share some of the chapter headings of a third grade textbook from a Hindi series. The chapter headings translated into English read:

> The Dacoit Who Became a Brother
> Autobiography of a Tree
> India is my Home
> The Victory of *Dharma*
> The Sage of Samaria
> A Trip to the Zoo
> Water Is Life
> Glimpses into the Cities of Rajasthan
> Alexander and King Porus
> Patriot and Freedom Fighter Chandrashekhar

These were a few chapters among several others in the book (Kapoor, 2002). From the titles it is clear that the third graders would be studying stories about India as well as other countries, about contemporary as well as historical events, about science and social studies as well as Indian philo-sophical concepts. For instance, the chapter on the Autobiography of a Tree begins with the information that in India's ancient books on *Dharma* the value of trees has been given much importance. It goes on to describe the tree as a protector of beasts and humans, and a provider of shade, food, and fuel. Large sections of the chapter are devoted to how ancient *ashrams*, or the very first schools, were set up under the shade of a tree where the *Guru* would educate his students; and how *Lord Krishna* would lean against the trunk of a tree and melodiously play his flute. This eclectic mix of chapters was a clear example of how Indian and Western, ancient and modern, religious and scientific elements were all pulled together in this Hindi text, which would have constituted just one part of the larger third grade curriculum in a school.

I would again like to reiterate that there is great diversity in the context of schooling for young children in India, and a study of poor, rural, or government schools would reveal a very different picture. Private, urban schools in India have been constructed within a specific middle-class cultural context that includes a strong interaction between Indian and Euro-American ideas. Today, the influence of the American media and toy industry is powerful, and urban children in India are exposed to a different worldview through the Western occupation of cyberspace, media, entertainment, and business in the name of globalization. The Indian teacher in this context appears to enact a curriculum that attempts to do the following: to balance the values underlying Indian philosophy and the values depicted by Western media; to balance tradition and so-called progressivism; to balance academic learning with experiential learning; and to balance the Indian colonial and postcolonial elements in her practice. The language of the discourse of urban Indian early childhood education, and the nature of this complex curriculum defies definition and can only be understood within the locus of the system of intersecting values and beliefs that is constantly and profoundly influencing daily life inside and outside the Indian classroom.

The educational environment in urban India is one that is extremely competitive with high academic standards, high stakes assessment, textbook-driven examinations, and a tremendous pressure on students and teachers to produce high academic results. These attitudes prevail even at the early childhood levels. The critical finding in this study seemed to be that teachers recognized the importance of working with the prescribed academic syllabus and helping students develop the skills to succeed in this competitive environment. But they also recognized the importance of teaching the whole child and ensuring the development of social, emotional, and moral development. More importantly, teachers seemed to have the freedom to individually implement a more informal curriculum parallel to the rigid academic curriculum where issues in values, good attitudes, environmental protection, and diversity were being addressed. What was striking was not the evidence of three different ideological influences in the early childhood curriculum, but the unvoiced, underlying recognition by the teachers that certain aspects of each approach had a place in their classrooms and in the overall success of the child's educational experience. There seemed to exist a porous membrane around the concept of early childhood curriculum whereby ideas transcending diverse times and cultures could be selectively assimilated; and amoeba-like, the shape of the curriculum could shift to adapt to specific educational goals. The curriculum was neither purely academic nor child centered, but presented an early childhood knowledge base that had been socially constructed within a specific historical, political, and values context.

The educational outcome we should aim for is not tolerance or the celebration of difference, but more so the critical reexamination of difference, which will allow us to question our own systems and the implications they hold for us and other people (Burbules, 1997). In relating this study to the larger debates within Western academia, the awareness of this "different" approach to early education in urban schools in India can perhaps lead to a deepening of our own understanding of the childhood curriculum as being far more than only a printed model or a scripted guide for teachers, or only a loosely flowing open-ended emergent experience for children, but rather a socially constructed, constantly "becoming" complex phenomenon combining the academic, creative, and socioemotional development of children. It would be "becoming" in the sense of constantly shifting and evolving, and "becoming" also in the sense of suiting the changing patterns in the needs and expectations of families at the local level because, ultimately, a curriculum is really a reflection of who a people are. The debate on which is a more appropriate curriculum for young children, academic or child centered, has been ongoing for over 100 years and still remains unresolved. It is critical to stop asking either/or questions, stop looking for one right answer, and, more importantly, stop feeling compelled to side with only one particular approach. Instead it is more important to acknowledge the fact that both experiences are equally important and begin to feel comfortable with the idea of a gray zone. There is no one right early childhood discourse for all societies and children everywhere at all times, and educators need to conceptualize an integrated curriculum as developing out of the expectations, aspirations, and struggles of those who are experiencing the curriculum most intimately. My hope is for this book to provide a different and alternative perspective on how the complexities of a curriculum can be variously confronted in different cultural contexts.

Chapter 10

Aligning Teacher Education and Early Childhood Practice in Urban India: Balancing Vygotsky and the *Veda*

The concept of teacher education as a formal discipline was introduced in India by the Europeans when Danish missionaries established the first institute to train teachers in 1787. Since then subsequent degree programs in teacher education in India have been defined largely by European standards. To get an idea of the varying approaches that have been adopted by Euro-American teacher education colleges over the years, I briefly review some of the more prominent models by highlighting their basic premises.

Varying Approaches in Teacher Education

For the longest time, teacher education in Europe and the United States had been considered as the teaching and learning of a technical skill, and classrooms as places to practice these specially learnt techniques, without any regard to the complexities of students' lives (Delpit, 1995; Simon, 1992). In light of this, teacher education has been more commonly known as teacher training and consequently has sought to develop in teacher candidates a basic teaching competence and to train them to precisely implement specific teaching strategies. This perspective links teaching as a technical skill to the positivist tradition in education (Bartolome, 1994). Thus traditional teacher education has been based on a training model in which the college teaches the teacher candidates theory and the methods; the schools and actual classroom settings are where the candidates practice that knowledge; and the candidates themselves provide the individual effort that demonstrates the application of that knowledge (Wideen et al., 1998).

The stage theory model of teacher development draws on the works of Kohlberg (1969) and Piaget (1965/1932), and holds that teachers learn how to teach in developmental stages. Explaining this perspective, Tischer and Wideen (1990) suggest that teachers focus on different concerns as they move from one stage to the next. Starting with concerns about their own survival as beginning teachers, next they focus on their concerns regarding effective teaching strategies, and then finally they begin to focus on their students and learner outcomes. This research drives many teacher education programs in the United States. However, both these approaches, based on the training model theory and the stage theory, tend to disregard the idea of teachers bringing to the field a wealth of personal experiences, maturity, and values. The truth is that teachers' practices are embedded in the complexities of their own experiences and values, which need to be taken into consideration by teacher education programs (Phelan, 1997).

A constructivist model or approach to teacher education helps teachers learn new ways of thinking about teaching and learning, acquire better knowledge of different subject matters, form communities to allow for shared understanding and dialogue, and develop democratic structures of authority in schools. Britzman, Dippo, Searle, and Pitt (1997) emphasize that in order for teaching to be considered more as a relationship between pedagogical strategies and factors such as language, culture, identity, and knowledge construction, it has to be reconceptualized as a problem of learning where learning is a sociocultural construction. This kind of teacher education will be driven by experience, dialogue, and discourse. The language of experience includes how an event, condition, or action becomes narrated, rethought, and then becomes an occasion for learning (Moll, 1990; Vygotsky, 1978). Such a language of experience is invaluable in the discourse of teacher education. Further, constructivist pedagogy incorporates two premises, namely (1) the starting point of instruction must be the knowledge, attitudes, and interests that teacher candidates bring to the learning situation and (2) instruction should be designed such that the experiences provided can interact with the characteristics of teacher candidates who can then construct their own understanding (Howe & Berv, 2000).

Teacher education programs have also increasingly turned to the reflective model and the belief that thoughtful teachers are more effective teachers. The foundations for reflective thought in teaching lie in the works of Dewey (1933) and Schon (1987). According to Sparks-Langer and Colton (1991), there are three approaches to reflection: (1) cognitive, in which teachers use their pedagogical knowledge and content knowledge in planning and decision-making; (2) critical, in which teachers examine the moral and ethical aspects of a problem situation and its proposed solutions; and (3) narrative, in which teachers draw on their own personal experiences to make judgments and decisions. Teachers encounter a state of uncertainty or a problem, use their knowledge and experience to contextualize the problem, and through careful thought devise a solution to the problem. Thus reflective teaching addresses both personal experiences as well as formal pedagogical knowledge. In a similar vein, Tauer and Tate (1998) urge teacher educators to become reflective practitioners as they examine present experiences in the light of past ones.

An approach to teacher education that has been under much discussion is the model based on multicultural education, or more specifically, multicultural social reconstructionist education (MCSR), which was suggested by Sleeter and Grant (1994). This itself, as an outgrowth of critical theory, and the works of Giroux (1989), McLaren (1994), and others addresses issues such as cultural analysis and the understanding of how individuals construct meaning through social and cultural interactions. Such an approach encourages a pedagogy that supports a context in which diversity is acknowledged and valued. This discourse urges teacher educators to explore the relationships among diversity, systemic oppression, and pedagogy. Most proponents of MCSR agree that central to this approach is critical reflection by prospective teachers on their own experiences (Farber, 1995; Martin, 1995). The work of Allsup (1995) highlights the important fact that when teacher education approaches examine issues of diversity but embrace and validate the dominant culture, it immediately marginalizes any values that are in opposition to the dominant values. Martin (1995) notes that two major obstacles that face MCSR are: (1) political and institutional guidelines, state mandates and policies, teacher certification requirements, course design, and so forth; and (2) the challenge to change individual worldviews, attitudes, and behaviors of teachers and student teachers. bell hooks (1989) in her work discusses the challenges faced by prospective teachers who have no exposure to the issues of diversity and who are unwilling to admit that their minds have been conditioned to accept the domination of others.

A large body of recent research on teacher education has shifted the focus from school effectiveness to teacher effectiveness. This approach recognizes that effective teaching competencies necessarily include attitudes, dispositions, and interpersonal skills that teachers exhibit while interacting on a daily basis with students, families, school administrators, and other teachers. There has, thus, been an increasing interest in the importance of the socialization of teachers, or in other words, the processes whereby individuals acquire the values, skills, knowledge, and the culture of teaching. Terms such as social skills, interpersonal skills, emotional intelligence, social competence, social cognition, and so forth have gained prominence in teacher education discourse. This teacher competence within the context of the workplace is a complex phenomenon and involves ecological, genetic, and action-related processes (Brouwer & Korthagen, 2005), or teachers' backgrounds, local contexts, and state policy environments (Achinstein, Ogawa, & Spieglman, 2004). Currently, there appear to be few systematic efforts to explicitly teach teachers how to develop strategies for effective social interaction. How teachers react to different social situations is largely attributed to individual differences based on personality, background, or style (Mills, 2003).

In conclusion, one can summarize that a teacher education program that does not address issues of implicit beliefs, diversity of experiences, and significance of language and culture of all its participants, and a program that does not address the imposition of dominant values that are rooted in hierarchy and power, is not taking into consideration vital factors that shape pedagogy, and is thus doing an injustice to the preparation of future teachers. These issues become all the more critical and complex in teacher education programs that prepare teachers in nations that were once colonized. Viewed through the lens of postcolonial theory, the complexities of cultural hybridization need to be recognized and the effectiveness of teacher education programs in formerly colonized nations such as India may be enhanced if they offer a formal study of ideas based on an understanding of child development and educational practices that make sense within the local context.

Figure 10.1 presents a summary of the models of teacher education discussed in this section, along with their objectives and underlying ideas.

Overview of Teacher Education in India

A UNESCO survey from 1972 provides a brief history of teacher education in India and makes mention of the fact that the first teacher education institute in modern India was established by Danish missionaries in 1787 (Teacher Education Asia, 1972). A central school for the professional education of teachers was established in Chennai (formerly known as Madras) in 1826. In 1901, there were 155 teacher education institutes in the country to prepare teachers for the primary level. A recent survey indicates that by 1998 India had 848 recognized university systems with a teacher training

Teacher education model	Objectives and underlying beliefs
Training model	Teaching is viewed as a technical skill and individuals are trained for the precise implementation of specific teaching techniques and strategies.
Stage theory model	Teachers focus on different concerns as they move from one stage to another in their development as teachers: from initial survival to teaching strategies to learner outcomes.
Constructivist model	Teaching and learning are viewed as sociocultural constructivist processes drawing upon experience, dialogue, and discourse.
Reflective teaching model	Teaching involves creating strategies and making decisions by using cognitive, critical, and narrative processes.
Multicultural and social reconstructionist model	Teaching involves a pedagogy that supports diversity and cultural analysis, and opposes systemic oppression, acknowledging that individuals construct meaning through social and cultural interactions.
Teacher socialization model	Effective teaching competencies such as dispositions, attitudes, and interpersonal skills are acknowledged and addressed.
Postcolonial teacher education model	Recognizes the notion of cultural hybridity and the transaction between local and "Western" ideas and practices, and designs a program of study that offers various perspectives in the formal curriculum.

Figure 10.1   Models of teacher education and their underlying beliefs

component. A 2001 report on the National Council of Teacher Education (NCTE) mentions that there are about 1,300 teacher education institutions for elementary teachers, and nearly 700 colleges of education/university departments preparing teachers for secondary schools (www.ncte-in.org). In all there are about 2,600 teacher training institutions in both private and government sectors that provide pre-service and in-service training to teachers and teacher educators. Data available from India's Ministry of Education indicates that there were about 4.3 million teachers in India during 1994–1995 working at different levels of schooling. That number will, of course, be much higher now.

Since India's independence from British colonialism in 1947, several attempts have been made to change the system of teacher education that was inherited from the European rulers, but despite change efforts the system still continues to function in largely the same way in terms of principles, approaches, and content of the teacher education curricula. Over the years India has developed a multitier infrastructure for teacher education and the existing programs are generally based upon the teacher education curriculum framework presented by the NCTE in 1978. A fully qualified primary teacher is required to have 12 years of school education, followed by 2 years of primary teacher training. These are the minimum requirements as recommended by the National Policy on Education report of 1986, but the practices differ from state to state. Most private schools require a college degree in addition to the teacher training experiences. The majority of primary teacher education institutions are under the administration of the government. NTT institutions offer diploma courses for early childhood teachers. The Ministry of Education provides a description of some of the leading teacher education organizations. At the national level, The National Council of Educational Research and Training (NCERT), established in 1961, is responsible for designing exemplar instructional material on teacher education and has provided training through innovative programs. The NCTE, set up under a Parliamentary Act, has focused on the planning and coordinating of the development of teacher education. The Indira Gandhi National Open University has a School of Education that offers teacher education as a distance learning program. SCERTs were set up in 20 states, and State Institutes of Education (SIE) were set up in 9 other states as counterparts to NCERT to provide direction and leadership to reforms in school education and teacher education. Under all of these organizations, there are elementary teacher education institutions at the state levels and District Institutes of Education and Training (DIET) at the district levels, which are continuously upgraded under a Centrally Sponsored Scheme (CSS). The Restructuring and Reorganization of Teacher Education, a CSS, was set up in 1987 to ensure a sound institutional infrastructure, to maintain a technical and academic resource base, and to continually evaluate and upgrade the knowledge, competence, and pedagogical skills of elementary school teachers in the country. The NCTE was established by the Government of India to ensure the coordinated development of the teacher education system, and to regulate and maintain the norms and standards of teacher education. After a process of nationwide consultations with teacher educators, teachers, and educational philosophers, a new curriculum framework for teacher education was developed by NCTE and presented at a national seminar in 1998. The primary aim was to free teacher educators and teachers from rigidly prescribed instructional and assessment strategies.

There has been limited research available on specific teacher education college curriculum in India. Singh (1980) has surveyed aspects of teacher education in India in an attempt to pinpoint areas that need improvements. Most teacher education programs focus on how best a teacher may be prepared to impart skills and knowledge to the learner. Teaching techniques and strategies have been heavily imported from the United States and United Kingdom and are given prominence in teacher education. He also notes that there is a general lack of motivation and stimulation amongst teachers. One finding is that the actual needs of the teacher candidates are not considered, and there is no connection between who the teachers are as human beings and professionals. This might be interpreted as a gap between teachers' personal and professional beliefs and experiences. This may also be in part because of the strong focus on Euro-American ideas and techniques, and because the teacher candidates are not exposed to the understanding of the social and historical evolution of education in India, and to such ideas as the differences between wisdom and knowledge as understood within an Indian frame of mind, or to the understanding of the different ways of knowing as outlined in Indian texts and practiced through local values and traditions. One interesting example is that Ancient Indian texts recognize seven means of acquiring knowledge—through *pratyasa* (perception), *anumana* (that which can be inferred), *shabda* (through verbal testimony), *upamana* (comparison between two cognitions), *arthapatti* (presumption), *sambhava* (cognizance of a part from the knowledge of a whole), *and aitibya* (succession of rumors and oral tradition). This is explained in the educational philosophy of the *Veda*. If such subjects could be included in teacher education programs, it would, no doubt, increase not only the interest levels of teachers but would also deepen their understanding of students. Singh notes also that although research has been conducted by the NCERT in India, areas of teacher education that may not have been touched upon are philosophy of education, history of education, and comparative education.

Another survey compiled by Srivastava and Bose (1978) presents an examination of the curricula in teacher training colleges in ten cities in India—Ajmer, Aligarh, Allahabad, Bhopal, Gwalior, Jodhpur, Lukhnow, Meerut, New Delhi, and Pratapgarh. Both the researchers were professors of education at The Central Institute of Education in New Delhi and conducted the study as part of a national project titled *Evaluation of practice teaching in teacher training institutions*. The study comprised the documentation of syllabi and curriculum; analysis of evaluation procedures; and interviews with principals, teachers, and students at each college. The objectives of teacher education were broadly categorized as: understanding the principles of education; understanding the growth and development of the individual; understanding how an individual learns; developing and practicing the skill of teaching; understanding the psychology of learners and techniques to evaluate learning; and considering the participation of students in cocurricular activities to promote social development and a balanced personality. An interesting finding from this study indicated the relative importance given to three areas in teacher education—pedagogical knowledge (20 percent); working with the community in rural and urban areas (20 percent); and subject matter, methodology, and practice teaching (60 percent). However, in describing pedagogical knowledge, colleges consistently emphasized knowledge of educational psychology, whereas the philosophy of education is conspicuous in its absence.

Teacher education in India has been largely conventional in nature and purpose. The integration of theory and practice is inadequate, teachers are prepared in competencies and skills that are technical in nature, they are not adequately exposed to the latest developments in education and research, and, further, are not prepared to be reflective thinkers who can critique educational research. I have provided only a brief overview of the system of teacher education in India for the reader to get some idea of the infrastructure in place. A detailed examination of the same is beyond the scope of this book and the reader may access other literature and websites in order to do so.

Visits to Two Teacher Education Colleges

The two teacher education colleges that I have referred to in my study are both part of University of Delhi. The university system in itself is a government organization consisting of about 80 multidisciplinary colleges located in the city of Delhi, which is inclusive of both New Delhi and Old Delhi. I refer to the two colleges as College A and College B.

College A was established in the early 1930s to provide opportunities for women to become professionals and give them status in the academic world. It was started as part of the national independence movement by a group of very dedicated women who were deeply committed to the cause, and offered programs in Home Science and Home Economics. With time, as the institution flourished it acquired a scientific flavor and is currently based on four disciplines, namely Child Development, Nutrition, Textiles, and Resource Management with an extension in Community Services. It is thus a Home Science institute but offers the Bachelor of Education (B.Ed.) program as a graduate degree after the completion of the undergraduate degree. The institution also runs a children's school but this lab school is affiliated with the Department of Child Development in the college and not with the B.Ed. program. This highlights the strong divide that has existed in India between the field known as Education, which addresses secondary school issues, and the field known as Child Development, which has historically addressed issues in early childhood education.

When I arrived at College A for my first visit, I found it to be a bright, open place with low brick buildings that were surrounded by a brick wall with several gates. I saw several young women walking about on the campus and I assumed they were students at the college. I walked toward the child development center, a school for young children, where I had been told that I would find Ashwini. I spotted her in the sun-filled open courtyard with a group of young women. They were all sitting on *moodhas* (wicker or cane stools) placed in a circle, and were talking and discussing something in an animated manner. Later, I met with Ashwini in her office to interview her. The office windows and doors were open. The door overlooked the courtyard below, which was surrounded by the nursery school classrooms. The office windows opened out toward the exterior of the building, which faced a large and busy road. As we spoke, sounds of children laughing and playing mingled with sounds of passing cars and buses and other sounds of the street. There was a feeling of warmth and comfort.

College B was established at the time of India's independence in 1947 and has been one of the nation's premier institutions for professional learning and research in Education. The college was envisioned as an institution that would evolve into a research center to address educational problems of a new India, prepare teachers to be model teachers, and also be a model for other training institutions of the country. College B offered the following academic programs: (1) a full-time teacher education program, B.Ed., offered for one year after a Bachelor's or Master's degree in a content area to prepare teachers for middle and secondary school teaching; (2) a recently established full-time teacher preparation program, B.El.Ed., offered after 12 years of schooling to prepare teachers for elementary school teaching; (3) a full-time advanced program in Education leading to an M.Ed. degree; (4) a part-time advanced program leading to an M.Ed. degree for practitioners; (5) a full time predoctoral research program for a duration of one year leading to an M.Phil. degree; (6) a doctoral research program leading to a Ph.D. degree; and (7) other in-service programs for teachers and practitioners.

College B consisted of low brick buildings spread out over an extensive campus. The bricks were painted a garish terracotta. There was silence in the hallways, which were wide, open, and had several arches all around the perimeter. This verandah-like feel was a remnant of the British Raj and reminded me of India's historical colonial past. Some spaces in the verandahs were lit by bright sunlight wherever the sun's warm rays could find an open archway. The rest of the hallway area was dark and cold on this January morning. It seemed there was no electrical lighting to light the dark corridors. In the almost tomb-like silence, hushed voices suggested the presence of human beings. Bulletin boards lined the walls of the corridors inside the buildings and were covered with charts and posters. The students at this college appeared to be grouped into "Houses" named after some of India's national leaders. The charts were beautifully inscribed in script and calligraphy with poems, noble thoughts, and inspirational sayings. As I read these messages I got the feeling as though the teacher candidates had chosen a life of austerity, sacrifice, and nonworldliness. To me this image of the teacher was very remote from the energy, diversity, and complexity of the Indian world outside these walls. For some reason the notion of an "ivory tower" came to me as I stood in the grave silence of this academic world that prepared individuals for a highly noble vocation. But how did this image relate to the image of the early childhood teacher who was expected to go into classrooms with a feeling of joy, enthusiasm, wonder, and love for the children? This contradiction puzzled and haunted me for quite a while.

Later, I waited in a room for the professor with whom I was scheduled to meet. I looked around and guessed the space to be an office-cum-classroom. A chart on one wall displayed the phrase "problems of curriculum load." Listed under the heading were such words and phrases as: joyless learning; exam system; textbook as the truth; observation discouraged; structure of the syllabus; starting early; teaching everything; not just an urban problem; and so forth. I would have loved to be a fly on the wall during a class discussion as each one of these issues was explored in detail. The list sounded critically reflective of many of the current problems of education in India. It made me think that, perhaps, the right conversations were indeed taking place within teacher education classes. But the status quo still seemed to persist. So where was it exactly that change needed to be initiated?

I had now been waiting for an hour and a half and I was quite sure that I had not been mistaken in my appointment for this morning. I spoke to the clerk and the librarian in the small adjoining library and they informed me that it was a day when most professors were supervising student teachers in the field. I walked up to speak with an elderly looking gentleman and introduced myself, explaining that I needed to speak with any one of the professors at this institution. He could not help and I got an overall impression of unhelpfulness as I left the campus.

Turning back to the perspectives of the teachers I interviewed, I recalled how Charu had described her teacher college experience as a student: "It was like being lost in an unknown place. That was my initial reaction, I remember the first day . . . I thought My God, how will I ever go through one year in this place! It's so impersonal and so cold. Of course later on, one makes friends and so you just go through the whole year. But basically you just attend classes because you have to. It's not because you're enjoying it or you want to learn. That's how I felt. Very cold atmosphere—there was no warmth, there was no friendliness." Malvika also shared her memory of the course content and pedagogy of her teacher education college curriculum: "basically we had Psychology where we studied child psychology . . . we had Piaget and all, and we studied the basic developmental theories . . . we also studied various personality traits, personality defects, problems of adolescence . . . But then we were never given a chance to relate all these things in practice. It was only for the exam time that we had to mug it up and get marks [scores]. That was the only thing it was related to . . . And we talked about basically western philosophers like Rousseau and all that. We talked a lot about them though it wasn't easy . . ."

The course content and pedagogy to this day remains very colonial in nature marked by traditionally delivered lectures, where the professor is the dominant speaker in the classroom; a strong focus on Euro-American theories of learning and development; a strong focus on stage theories of cognitive development, the history of education in Europe and America, the educational philosophies of Rousseau, Dewey, and so forth. Highlighted below are a few topics, theorists, and philosophers mentioned in the syllabi of some of the courses offered in a four-year Bachelor of Elementary Education (B.El.Ed) degree program: (a) child development theories included those of Piaget, Erikson, and Kohlberg; (b) the ideas of Piaget, Bruner, and Vygotsky were included in the context of the study of logico-mathematical education; (c) the study of curriculum included ideas of Dewey, the models of Tyler and Bloom, the child-centered and experience-based curriculum, among others; (d) the educational philosophers mentioned were Rousseau, Pestalozzi, Montessori, Dewey, Susan Isaacs as some of the Western philosophers, and Tagore, Gandhi, and Krishnamurti as the Indian philosophers; (e) the socio-cultural, socioeconomic, and political issues studied were located in the time after India's independence and in contemporary Indian society; (f) and some of the other topics included in the curriculum were Language development and acquisition, Linguistics, Gender and Schooling, Special Education, School Planning and Management, Human Relations and Communications. In addition, a large portion of the course work was assigned to the methodology and pedagogy of science, math, English, social science, environmental science, and so forth; The content areas studied included math, biology, history, chemistry, physics, political science, geography, economics, English, and Hindi. This brief glimpse into some of the coursework included in the syllabus of one of the teacher education colleges provides some idea of the content of the formal course work studied.

As I described in chapter 7, the teacher participants expressed that the most important things they learnt from their teacher education programs were technical in nature such as lesson planning, making/using audiovisual aids, class control, time management, blackboard writing skills, maintaining registers and logs. When I asked Charu about the theories and whether she remembered theories she had learned in the Psychology of Education course, she said emphatically, "No, I don't! They were horrible sounding names which we had never heard before and we just mugged [memorized] it and reproduced it . . . As it is we had such large class numbers, and it was like a university lecture—they [the professors] just came and gave us what they knew and then they went off. It should have been done in smaller groups, with interaction where the trainees [teacher candidates] could question the teachers, and they [teachers] could give more practical feedback."

I had been able to schedule interviews with teacher educators from these colleges and had very interesting conversations with both Ashwini and Nagma. My meeting with Nagma helped me gain valuable insights into the nature of teacher education experiences as she reflected on the gap between teacher preparation and teacher practice. Not surprisingly, the problem was not localized in the teacher education program but was a more systemic one that had its roots in the overall structure of the educational system in India. Teachers were somewhat prepared and then put into schools where the tightly prescribed curriculum drives their practice. Of most importance in the system are the "marks" or test scores. This importance filters down from grade twelve to the Nursery classes. Within this system, in order to achieve academic excellence in their classrooms of at least 40 students, the skills and competencies that teachers need to be taught most are: (1) knowledge expertise and content-based teaching for students to score high in the national tests and examinations; and (2) classroom management and discipline. Even if teachers were prepared to be reflective practitioners there was little scope for them to do this in an actual classroom because of the intense competitiveness amongst schools. This competitiveness occurred to the extent that many school administrators were unwilling to share their professional or curriculum development experiences with other schools, and a one-upmanship spirit seemed to pervade the private school circles.

Questioning the Status Quo of the Teaching and Learning of Teacher Education

Based upon my own experiences and assumptions, I could summarize the following: (1) in education, formal theory is often disconnected from actual practice especially when multiple cultures are involved; (2) teachers' practice is largely influenced by their own construction of knowledge through social, cultural, historical, and personal experiences; (3) the nature of these influences on teaching practice varies greatly with the diverse cultures and societies that teachers belong to; (4) the powerful social, cultural, and historical forces in Indian society need to be acknowledged and contextualized in teacher education in India; and (5) the voice of the early childhood teacher in India needs to be heard.

Based upon the analysis of the data for this particular study the following conclusions may reasonably be drawn:

- There appear to be three distinct influences that act upon the relationship between the preparation and classroom practice of early childhood teachers in private schools in New Delhi: Indian philosophy, British colonialism, and American progressivism. The teachers are thus engaged in an intercultural negotiation and transaction of Indian and Western ideas and beliefs.
- The aims of education are largely consistent between the home and school of the child. Families and teachers work closely toward achieving those aims which include reinforcing the right values and attitudes, developing the intellect and the ability to think, acknowledging and respecting cultural diversity, and developing academic proficiency.
- The consistency between home and school is also manifest in the adult interactions that the child experiences with teachers in the school and with family at home. Adults tend to set clear and firm boundaries, offer honest praise and criticism, love the child, and usually try to model the behavior they want the child to learn.
- The prevailing image of the teacher in India in many ways mirrors the traditional, precolonial Indian teacher.
- Teachers' learning occurs primarily from informal influences within a sociocultural–historical constructivist framework, and which strongly informs their teaching practice.
- The gaps in teacher education in India are largely due to the fact that first, it is a colonial construct based primarily on a Euro/American design and discourse with imported theories of education; second, it is much too didactic and technical, failing to offer teachers experiences in sociocultural learning; and third, the scope of student teaching is too limited and teachers rarely develop an adequate understanding of the challenges of actual classroom teaching.
- Three components are seen to comprise teaching practice in the lower grades: (1) the teaching of content knowledge and academics, which requires of them the knowledge and expertise in a content area obtained through college specialization of a major; (2) the teaching of values and attitudes viewed as appropriate, which requires of them the knowledge and experience of sociocultural–historical activities through interactions with family, extended family, friends, and other social elements; (3) teaching to the test and examination, which requires of them the knowledge and expertise of testing, designing exams, evaluation, grading, and lesson planning, which they acquire from their formal teacher education curriculum. It might be relevant

to remind the reader once again of the high value placed by Indians on knowledge, learning, and education in general, and of a strong emphasis on academic excellence in the highly competitive race for educational and job opportunities.

Aligning Teacher Preparation and Practice

I would like to now focus specifically on some of the gaps that have been experienced in teacher education and examine how the processes of teacher preparation and teacher practice can be brought closer together. In being more closely aligned they might succeed in better informing each other instead of raising contradictions, and thus develop in teachers a higher level of confidence in their own classroom practice.

Tighter Connections between Teacher Education and Teachers' Beliefs

The process of sociocultural constructivism as being an important consideration for the learning process was discussed in chapter 1 with reference to children's learning. In chapter 7, I developed the argument that teachers also, in learning to teach, engaged in a process of sociocultural constructivism. The fact that teachers as students of education construct knowledge through social and cultural experiences has been supported by several studies in the past (Clandinin & Connelly, 1992; Delpit, 1995; Elbaz, 1981; Jalongo & Isenberg, 1995; Williams, 1996) and it has been recommended that teachers should be able to acknowledge and name these experiences in their teacher education programs. Knowledge of educational theories at variance with the knowledge teachers have constructed through their own experiences will tend to remain absent from their practice. Stott and Bowman (1996) note that theories cannot be universalized, and in order to prevent a discrepancy between formal theory and a person's experience, it is important to bear in mind that (1) children do not exist outside of a social context, and their development can only be understood within cultural and historical conditions, and (2) values and beliefs underlie all theory and practice. Additionally, learning to teach is also learning about self, and the challenge is how to make teacher candidates more aware of their prior beliefs, and move teacher education toward a deeper understanding of the socio-constructive nature of the cognitive development of teachers (Schoonmaker, 2003). Teachers' practices are embedded in the complexities of the knowledge of their own experiences and values that need to be taken into consideration by teacher education programs (Phelan, 1997). A socio-constructivist approach in teacher education offers teachers the option to learn new ways of thinking about teaching and learning. As mentioned earlier in the chapter, this kind of teacher education is driven by experience, dialogue and discourse, and this language of experience includes how an event or condition or action becomes narrated, rethought, and then becomes an occasion for learning (Moll, 1990; Vygotsky, 1978). In this study, the language of experience may very well be the Indian ways of thinking and living. Subsequently, it stands to reason then that this language should have a prominent place in the discourse of teacher education in the Indian context.

When educational theories do not relate to the overarching philosophy that shapes the belief and value systems of a particular society, conflicts occur in pedagogy reflecting a dissonance between formal theory and practice. Research on special education in South Asia refers to the more than 2,000-year-old educational records of India and notes that the art of bringing up, handling, and playing with children was one of the recognized skills appearing in the educational curricula of Ancient India, and offers strong recommendations for the development of culturally responsive programs in India (Miles, 1997; Srinivasan & Karlan, 1997). This might also explain the limited success of concentrated efforts to develop child-centered approaches in Indian primary schools over the last two decades or so; unless adequate attention is paid to underlying concepts and philosophical assumptions, appropriate educational change will not happen. Thus, when teachers study theories based largely on Freud's psychoanalytical development, Piaget's cognitive constructivism, Erikson's psychosocial development, Skinner's instructional models based on behaviorism, and other Euro-American psychologies and theories in their teacher education program but are not introduced to the philosophies and psychologies that shape the belief system of the culture of their own society, then they are not being adequately prepared with a culturally relevant educational philosophy as the basis of their practice once they are in the classroom. In such cases, classroom practice is mostly informed by the teachers' own instincts and belief systems, which, in turn, draw from their collective cultural and historical experiences within their social environment. Teachers facing this dissonance between educational theory and their own practical knowledge may be likely to disregard altogether the educational theory studied, or may have low confidence in their own actions and in the underlying assumptions on which their own beliefs are based. This in turn can lead to some teachers who will be ineffective and who end up giving their students mixed messages about what is expected and acceptable in terms of behaviors, skills, and values (Katz, 1996; 1999).

When I asked Charu whether, upon graduating from her teacher education program, she felt prepared to teach the children the values, the skills, and the knowledge that she prioritized, she said:

> No, that came with experience . . . It took some months for me to feel confident and to know how to go about things . . . And the second year was better . . . Maybe they [the teacher education faculty] should have stressed more on that . . . So these things I feel should be incorporated in the syllabus during our teacher training . . . the important role you play in the life of children . . . the teacher training institutes . . . should inculcate how a teacher's moral responsibility is very very [double emphasis] important, you know . . . Besides [in addition to] her skills as a teacher, her sense of responsibility and that love for teaching is very important . . .

Malvika also felt the same about the B.Ed. curriculum: "we only talked about the academic part. How to teach them seasons, how to teach them this, how to teach them that. And hardly any stress is laid on value education or explaining our culture, or telling them about your culture. I think this curriculum also needs to be a little reformed . . . See, training [the B.Ed program] is one thing—you're being told how to handle forty or fifty students in a class . . . but your ways, your personal reflection will surely matter. And it's the personal experience . . . whatever you give them [the children] they carry it home—even your emotions . . ."

The dissonance between the preparation and the actual practice of teachers is a phenomenon that has been experienced globally, and the gap between theory and practice is found to exist in many countries and contexts (Brouwer & Korthagen, 2005; Korthagen & Kessels, 1999). Educational concepts developed during teacher education seem to fade away during field experiences and there is a distinct lack of transfer from teacher education to practice (Feiman-Nemser, 2001; Wideen et al., 1998; Zeichner & Tabachnick, 1981). Novice teachers come to learn more from their experienced colleagues at work whom they view as role models, and feel less prepared by their teacher educators. One of the chief reasons for this is the recognition that teachers' personal backgrounds, prior beliefs, and practical knowledge fail to be acknowledged sufficiently in their teacher education programs (Clandinin, 1986; Jalongo & Isenberg, 1995; Spodek, 1988; Williams, 1996). This becomes more of an issue in a country like India, where historically there exists a large body of literature from ancient times that presents the theoretical foundations for a way of life. Ancient Indian texts such as the *Veda* and *Upansihad* not only present a written belief system that is actively practiced even today, but also provide clear guidelines for a diverse range of disciplines such as philosophy, religion, politics, law, education, pure sciences, and the liberal arts. It is relevant to again draw attention to the misconception that the Vedic texts are purely religious in nature. As is evident in the previous chapters, religion is one of several disciplines that is addressed in these texts. So the exclusion of a culturally appropriate Ancient Indian educational philosophy from the more colonially selected body of educational philosophies taught in teacher education programs in India would seem to be a major oversight. A concern that many school principals in India have is that teachers are not adequately prepared for teaching the lower grades. On the other hand, a concern that many teachers educators in India have is that teachers leave behind all the theory they have studied in the teacher education institutions: once they enter practice, this theory has no place in their daily classroom work. One chief reason for these disparities is that although teachers' formal training draws on Western constructs, their practice draws largely upon their socio-culturally constructed system of beliefs and practical knowledge, and this is not theoretically validated in their teacher education curriculum. For the teachers in India, the Western perception of children's learning, growth and development is a purely theoretical discourse that does not acknowledge the powerful impact that cumulative social, cultural, and historical experiences have on childrearing and education of young children in Indian society. These teachers are then expected to oversimplify that formal Western teacher education knowledge by directly translating it into their classroom practice. Therefore, if schools in India aspire to offer sustainable educational programs that are values based, respectful of diversity, and hold high academic standards, then teachers need a teacher education program that is based on a discourse that addresses all these elements. Such a discourse might include the local Indian beliefs or theory/philosophy of education in conjunction with the more modern psychologies that inform a theory of learning. This is one way to ensure that teachers will feel validated, and comfortable and confident to facilitate children's learning and development in all dimensions—physical, intellectual, emotional, social, and spiritual.

Closing the Gap Through Student Teaching

One of the existing gaps between teacher education and practice was the disconnection between colleges of education and actual classrooms. But interestingly, and despite its inadequacy, the practical experience called practice teaching (student teaching) was perceived to be the most meaningful experience by the teachers I interviewed. Although they faced many tensions during student teaching, it brought them into direct contact with real students and gave them a more tangible sense of being a teacher.

Charu remembered "the best part was the practice teaching" because that made more "sense" as compared to the "boring" lectures on theories, which "mostly focused on . . . theoretical things of education—their education acts and all their policies which basically we should know broadly, but we're not concerned with that in our day-to-day teaching." But even the preparation for practice teaching did not allow for any discussions on what the student teachers might encounter in terms of the socio-emotional dynamics of any classroom. When I asked her if any connections had been explicitly made for her between the theory classes and practice teaching, she said:

> No, no, no no. We were just given a formal format on how to do a lesson plan . . . nobody was there to guide you at all. So initially it was very difficult, but as you did a few lessons in class it became a little easier . . . initially it was like being thrown into the water without knowing how to swim . . . You know, at the college if they had taken a model class [modeled a classroom situation], or given us these practical problems that you might face, then we would have been better prepared.

Charu also urged that teacher training institutes should emphasize that, apart from teaching skills a teacher's moral responsibility is very important, "you know, her sense of responsibility and that love for teaching . . . And probably they should do something like an Art of Living course, to reinforce positive thinking . . . If only the teacher herself is . . . positive, then the teacher can meet the children's needs . . ."

For Vasudha, student teaching had been a comfortable experience as she had been placed in the same school she herself had graduated from. Much of the curriculum was still the same and many members of the staff were familiar faces to her. She explained that the practice teaching, or student teaching block, occurred in two phases during the B.Ed. degree program, with 15 days of teaching lessons in a classroom in the first and second terms each.

Malvika had mixed experiences, but said: "I really really [double emphasis] enjoyed my practice teaching, in spite of all the stress that we got from the supervisors and teachers. I really enjoyed it . . . that was the first teaching experience, and with your new professional identity. . . ." In chapter 7 I have described a couple of incidents that were acutely distressing to her with regard to the rigid expectations from student teachers and the tensions they felt in trying to please the opposing viewpoints of their cooperating teachers and their supervisors. Malvika firmly believed teaching to be more than just the delivery of an academic lesson, and involving emotional and psychological interactions with the children. But these issues, she said, were not addressed at all during her teacher education program.

With regard to the nature of teacher education, the findings clearly indicated the following: (a) much of the theory in the course work was unrelated to the local culture and actual classroom situations; (b) many teachers were likely to experience a high level of stress during their student teaching due to differences in the expectations of the classroom cooperating teachers and the student teaching supervisors; (c) teacher education seemed to be focused more on preparing teachers to develop academic proficiency in students rather than teaching the whole child, and thus there was no discussion on the moral responsibilities of the teacher; (d) the socio-emotional environment in some teacher education colleges was lacking in warmth, personal attention, and collaborative work; and (e) there was insufficient recognition of who the teachers were as human beings, and how their socially and culturally constructed implicit belief systems determined their classroom practice.

In light of these findings, I want to highlight the strengthening of the following connections between teacher education programs and classrooms in schools during both teachers' preparation time and subsequent to their graduation:

- Student teaching needs to be more intensive and extensive. Student teachers would benefit with more time spent in the classrooms, not just delivering an occasional lesson, but also observing more experienced teachers at work and then taking on more responsibility.
- Teaching strategies adopted by teacher educators need to reflect the desired student–teacher interactions expected in early childhood classrooms.
- Early childhood education as a field should be included in the domain of a university-related Bachelor of Education degree program.
- Active mentoring should be provided while teacher candidates are still students at the college of education, as well as after they have graduated. A mentoring partnership may be established between colleges of education and schools, especially for first-year teachers.
- Frequent in-service workshops (as part of the teaching load of the faculty at the college of education) should be offered free of charge to be conducted in schools.
- Establishing a teachers' forum would allow teachers to share their experiences with student teachers as well as novice teachers.

But for Charu, the most important thing was: "there must be a paper [a required course] or, I feel, a topic or something . . . how a teacher's moral responsibility is very very important, you know. The trainee should develop morally also . . . probably they [college administrators] should also screen the trainees in such a way. Only if they feel they [the trainees] have a love or commitment to teaching should they be allowed to train for it, not just as a pastime . . . At least the good institutes can start with those steps . . ."

Defining Teacher Education Theory as a Postcolonial Venture

A contribution of this study was to provide an examination of one aspect of the Indian urban educational system within a sociocultural–historical constructivist and postcolonial paradigm. Although India has existed as a great civilization for thousands of years, it is a relatively young nation having at last freed herself of foreign rulers only in 1947. The last five decades have been characterized by the people of India searching for their identity and naming their "Indian-ness" after prolonged contact with European colonizers. As the immediate effects of colonialism have faded and Indians have realized their own worth, the desire to actively reconnect with their culture and heritage without having to completely forsake the benefits of colonial education has rapidly intensified. This desire has been seen to extend beyond Indians residing in India. With the start of the Internet, all across the Indian Diaspora there has been a surge of interest as people of Indian origin are rediscovering the roots of their cultural heritage. Postcolonial studies referring to India have been traditionally focused on the fields of literature, media, art, economics, history, and politics. Viruru (2001) has provided a postcolonial, ethnographic examination of an early childhood setting in Hyderabad in South India, which brings to the fore the voice of the young child in nursery school. It is my hope that this book will contribute to a direction in which the postcolonial discourse can be applied to childhood teacher education in India, thus enabling teachers in India to recognize their values and implicit beliefs, and to be able to articulate these beliefs as part of an international discourse on educational issues.

As discussed earlier, it is easy for educators in India to be enticed by the widely practiced child-centered developmental pedagogy of the West and accept that as a theoretical basis for teacher education in India. The discourse on DAP has been getting increasing attention among educators, school administrators, and policymakers in India. With the electronic revolution and the Internet as a channel of communication that has become accessible to almost everyone in urban India, the current practices in progressive early childhood education in countries such as the United States, the United Kingdom, Canada, Australia, and New Zealand are becoming available for educators in India to learn from. In addition, schools with resources are increasingly participating in international teacher exchange programs and conferences. As I have already done in chapter 6, I would like to caution again those who would strive to implement an imported Western notion of early childhood education in schools in New Delhi; there are many educators in India who buy into the desirability of the American progressive play-based pedagogical approach. It is vital to be mindful of what would be considered as being developmentally appropriate for children growing up in urban Indian society, and the skills that they would need to succeed and feel competent in that society. Postcolonial tensions are experienced as educators in India vacillate between ideas that are Indian and therefore "traditional," and ideas that are Western and therefore "modern." In order to bring about appropriate changes in schools and classrooms, it is necessary to consider carefully appropriate changes in teacher preparation programs.

I do not intend this discussion to be a debate of the relative value of Euro-American or Indian educational philosophy. As members of a formerly colonized society, Indians reflect both Western and Indian elements in their thinking and living. My objective is to reiterate the recommendation that an alternative to the current teacher education model is urgently required in India, a program that will clearly reflect a postcolonial curriculum which acknowledges the coexisting realities of both Indian and Western educational philosophies. Such a program, in my opinion, will better empower teachers in India to use a more culturally relevant pedagogy. Indian teachers have limited opportunity to reflect on their practice, and almost no professional forum to be able to consider educational aims in India within a sociocultural–historical framework, or to make sense of an educational theory that appears remote from their educational practice. Not only is it important for teachers in India to know about current educational research and discourses in various other nations, but it is also essential to equip Indian teachers with an educational discourse that will validate the underlying strengths of their own culture.

Just as life lived in the massive middle-class society of urban India reflects a complex intersection of traditional age-old values as well as more modern lifestyles, so also a multifaceted curriculum in many urban schools in India supports traditional ideas as well as modern ones. This consequently suggests a teacher preparation program that draws from both Indian and Western educational philosophies and psychologies. At present, there is no research study in India that has examined this issue. Teacher education programs in India can only be strengthened and energized if they formally include knowledge and a deeper understanding of:

- the local cultural and educational history;
- the development and evolution of education in India not just after independence but over the centuries;

- the origin of belief systems and traditions in India;
- learning styles that Indian children are comfortable with;
- developmental milestones that are prioritized for children growing up in India;
- the significance of the relationships between local childrearing practices and the roles that adults play in Indian society;
- appropriate social norms, interactional and behavioral modes commonly seen in India;
- speech and language patterns of Indians.

In addition, all of the above must be in conjunction with the study of current Euro-American child development theories and educational philosophies. This may point to a direction from which emerges a third form of formal educational theory for urban schools in India, supported by a postcolonial perspective, which would closely resemble the hybridized life styles that people lead in urban India. I find it critical to clarify here that my definition of "hybridized" would not imply an unrecognizable homogenous mix, but more like a matrix constituting ideas that are distinctly Indian, Western, or a combination.

Miles (1997) has recently made an attempt to draw attention to the vast heritage of educational ideas and practices of Ancient India, and his work mainly addresses research on special education; it can definitely be related to the examination of issues in regular education as well. According to him, personal differences in mental ability and attainment were first noted in the hymns of the *Rig Veda*, and he considers the *Panchatantra* to be the world's first special education text. If Ancient India had strategies for dealing with special education, then there had to have been an advanced and well-developed educational system in place. The cultural heritage of South Asia and the conceptual richness of its educational history that have been displaced with the inflow of Western educational ideas definitely need to be rediscovered. It is just such a plea that a postcolonial perspective on teacher education theory strives to address. The local indigenous voice of educational discourse in India that in many ways has been silenced due to the imposition of a more foreign dominant discourse needs to be heard. And at the very bottom of this hierarchy of educators' voices in India is the voice of the early childhood teacher, which needs to be heard the most.

I would like to further emphasize the significance of including a degree in early childhood education within the university-based B.Ed. programs; as well as the significance of basing teacher preparation in India on the realities of both Indian philosophy and teachers' implicit theories, as well as the more formal Western educational philosophies and psychologies. As Richardson (1994) notes, acknowledging personal and practical theories of teachers leads to an understanding and improvement of an immediate situation, whereas formal research contributes to a larger knowledge base. The first recommendation will help empower and prepare teachers more adequately for early childhood teaching in urban schools and ensure that school principals will not be forced to choose between B.Ed. college graduates who are trained to teach only the higher grades, and NTT candidates who learn the appropriate skills for early childhood teaching but are not college graduates. In terms of the second recommendation, theory that is based on Indian philosophy may work toward supporting a pedagogy that is relevant to a specific cultural context where schooling is taking place, whereas the theory that is based on Western philosophies and psychologies may work toward introducing an awareness of current global educational discourse and pedagogy. An intercultural teaching practice will emerge from an interfacing of the two perspectives, and may be one that is more culturally and developmentally appropriate in a given context supporting values-based modern lifestyles in urban India. Teachers will be better informed and more empowered to articulate their practice when the values and beliefs that they are imparting to the children are theoretically validated.

I make a strong recommendation for revising courses of teacher preparation in India. The current child development curriculum based on Piagetian and Eriksonian ideas needs to be revisited in light of the more culturally appropriate ideas that are permeating even Western early childhood discourse in Europe and the United States. Within the body of Western theories, a more prominent focus on Vygotsky's sociohistorical constructivism and other related theories is recommended, which would be more in line with the sociocultural–historical learning that teachers are already engaged in. The developmental theories from the Western discourse already included can be updated, with careful consideration given to a discussion of how these theories originated and evolved from overarching philosophical worldviews and the specific needs of societies in Europe and the United States. At the same time, a similar approach may be used in the presentation of the history and philosophy of Indian educational ideas and beliefs. A clear explanation must be provided about the parallels and contradictions between Indian and Western worldviews on the relationships between education, values, and social issues both in India and the West; the relationships between daily life, aims of education, and the overarching value system of a society; relationships between local childrearing practices and early education; the cultural constructions of children and childhood, of teachers and teaching; ways of knowing and understanding using the local discourse, and so forth. This is the kind of "social mooring" that Grant and Wieczorek (2000) consider as a tool to enlarge the frames of relevance in teacher education that will take into account existing connections between historical, social, and cultural perspectives, thus leading to a deeper understanding of the relationships between knowledge and power. This model of teacher education in India will not only enable teachers to better understand and name their own and their students' worldviews and help them recognize their own cultural values and biases (McAllister & Irvine, 2000), but also ease the tensions that the teachers face in their struggles to balance Indian and Western worldviews in their classrooms. The political context of teacher education must be able to address: (1) sociocultural realities in schools to which teachers are required to be morally and pedagogically responsive, and (2) government mandated educational policies, including tests and examinations, to which teachers must be legally responsive. Teacher education must therefore develop not only technical competence and knowledge of subject matter but also sociocultural competence in teachers (Moll & Arnot-Hopffer, 2005).

Recommendations for Future Research

Even as I raise the issue of change in early childhood education and teacher education in India, I raise a question that is even more critical: What kind of changes are appropriate for schools in India? The answer is certainly not moving toward being more compliant with the Western notion of progressive education in order to be on par with "developed" countries. The answer is also not in moving toward making education reflect only a philosophy from Ancient India. As is clear from this study, an understanding of the polymorphic nature of the Indian worldview today is crucial in determining how a corresponding educational approach can be effectively applied. In many instances, such an approach may already be in practice as teachers draw heavily upon their sociocultural–historical experiences that combine both Indian and Western ideas. But this practice needs to be formally acknowledged in the teacher education curriculum.

Future research may enable educators to respond to questions that remain unanswered in the present study with regard to teacher education programs as well as urban school practices. The only early childhood teacher educational program available in India for teachers who want to teach grades nursery and kindergarten is the NTT Diploma. This program is kept separate from all degrees offered in the field of education, although this might be what prepares early childhood teachers in the best practical way. Interview and observation-based research on early childhood teachers with the NTT background would provide valuable information, which could work toward designing a teacher education program in India specifically for a university-affiliated degree in early childhood education.

Future research areas could also include: (1) obtaining a deeper under-standing of why many school principals prefer to hire teachers who are college graduates rather than NTTs; (2) in negotiating ideas on the possi-bilities of implementing some form of child-centered and play-based cur-ricula in Indian schools, it would also be vital to have research information on the nature of children's play in India, and what it looks like in the classroom, in the school, and at home. Some questions to guide this research would be: What are adults' views on supervised and unsu-pervised play? Do children get the opportunity to engage in free play, socio-dramatic play when their school day is so limited (five hours) and so tightly scheduled? What materials do children use when they engage in free play and dramatic play in India? (3) parents' voices are always crucial in the understanding of the image of the child, the image of the teacher, and the aims of education. The research literature would be strengthened if that component were to be added and I highly recommend future inter-view-based studies with the parents of young children who study in urban schools in India; (4) another critical component in the understanding of all of the above issues are the children themselves. Additional research such as Viruru's (2001) ethnography would be invaluable to the field. Children's voices may be captured through interviews with young chil-dren in schools on their perspectives of work, play, freedom, academic expectations, values, and so forth.

Independent research in the field of early childhood education, which is not government sponsored, has been rare in India. I hope that further studies will be undertaken and that early childhood education in India can be explored more extensively. Constructing such a body of formal research will bring to the dominant discourse newer perspectives and different ideas on child development, childrearing, and childhood education in a non-Western nation, and one that is rapidly gaining prominence in today's world. A realization that is coming to the fore in recent years as India has emerged as a powerful player in the global arenas of commerce, technology, and economics is that one reason for this success is very likely the high value placed on education. This, as we know from our exploration of the historical, social, spiritual, and political influences on the Indian educational system, is not a new phenomenon but has been the case since ancient times and has been a value always prioritized in Indian society. It would be interesting to uncover and articulate other influences that may enable educators, policy-makers, and world organizations to understand and build upon, and thus strengthen the educational opportunities offered to children in India and elsewhere in the world.

Chapter 11

Reflections on the Process of Postcolonial Research in Early Education

While growing up in the 1970s, I attended an all girls Catholic school in New Delhi from kindergarten all the way through high school. Every morning, we boarded the school bus for the six-mile ride to school. I remember distinctly that as the bus neared the school it would pass a *yoga ashram*. From our seats on the bus we were able to peer over the lantana hedges and oleander bushes that surrounded the *ashram* to see dozens of people practicing yoga in the early hours of the morning. That smooth, green, grassy lawn with its human bodies engaged in a slow, silent, and ancient physical activity seemed so separated from the rest of the world that lay beyond the perennial hedge marking the boundaries of the institution. In contrast, within the confines of the boundaries of my own red, brick school building I was being exposed to a Eurocentric curriculum as I studied the psychology of Shylock, soaked in the social nuances and graces of the English gentry in their country cottages who came alive through the pages of *Pride and Prejudice*, read with awe the stories of Hannibal crossing the Alps, and of Romulus, Remus, and the founding of Rome, meticulously calculated the speed of a train traveling from Manchester to Birmingham, and played a noisy game of basketball on the asphalt court under a hot noonday sun. But that *yoga ashram* had always attracted my attention as a young schoolgirl. There was something hypnotic about it—the calm, the quiet, the controlled breathing and moving of human beings, the oneness of mind and body that appeared almost mystical. My father's childhood friend tried to explain the benefits of *yoga* to me as he himself practiced it every morning of his life. But much of my time then was spent on being a typical adolescent and teenager, reading a variety of books and comics replete with tales of love, war, and adventure revolving around Euro-American characters in Euro-American locales; and in wanting to do what was considered to be cool in the "West"—rock 'n roll, bell bottoms, Hollywood movies, and make-up. And yet, at home we also learned of "our" traditions as we participated in religious ceremonies and weddings, celebrated festivals, and viewed Bollywood movies. (Of course, it was not called Bollywood back then.) In the face of Westernization, there was wonder and appreciation for everything European and American, and yet at the same time, there was also a powerful consciousness about what was "ours" coming to us from our grandparents and other elders in the family. Life seemed to be a natural mix of things. Living with culturally diverse elements and learning to juggle Indian and Western identities was a way of being that we seemed to be comfortable with because that was the only way of being we had known.

The Research Site

I made the decision to conduct this study in early childhood education in a private school in New Delhi because it was a community in which I myself had lived and worked for several years as an adult. From my perspective as a researcher, it proved to be invaluable in being a native of Delhi and sharing the same background as the teaching community that was being studied in terms of being able to comprehend everything that was being said and being referred to, and being able to contextualize the participants' perspectives and experiences. On one hand, this can be a limitation since a high degree of familiarity can tend to influence and change the interpersonal relationship between the researcher and the participants. On the other hand, it certainly has the advantage of minimizing the misinterpretations that might arise due to differences in cultural worldviews and the complexities of multilingualism that are embedded in the Indian culture. More than anything else, it is the spoken language that presents the perfect example of cultural hybridization in India. In Delhi it is a curious mix of Hindi and English that is widely used colloquially, officially, commercially, and even in news broadcasts. Being familiar with the local language and vocabulary saved me an immense amount of time that otherwise would have been taken up by translations and interpretations. This would also have led to a more interrupted interview session. Some examples of local school jargon in English that were generously and frequently sprinkled into most of the conversations with the participants were:

- *copies,* used universally in India to refer to notebooks;
- *to pass out,* which means not to faint (!) but to graduate from an educational institution;
- *twelfthees,* a colloquial term for twelfth graders, or high school seniors;
- *bus books,* a term commonly used in many schools in New Delhi to refer to roll books used to mark the attendance of students who have been signed up to use the school bus for commuting between home and school;
- *cocurricular activities* is a term that is used universally in the language of education in India to refer to nonacademic activities such as sports, debates, dance, music, art, and so forth;
- *marks,* a term used universally in India to mean a student's scores or grades;
- *to do corrections,* a phrase that is universally understood in India as meaning to grade students' written assignments by the teacher;
- *timetable,* understood universally in India to mean schedule of classes for the day or week;
- *Prep class,* a term used to refer to the Kindergarten grade level in many private schools in New Delhi;
- *to study for a paper,* where "paper" refers to a specific course in an academic program offered by a college or university.

Being familiar with the language I was able to understand exactly what the participants meant when they used these and other terms and phrases, and did not have to stop to ask for translations during the course of my conversations and interview sessions. Most of the participants spoke in English throughout the interviews, although often their sentences were interspersed with Hindi words. There was a certain pattern that appeared in the speech in that many times sentences would be left incomplete as the speaker jumped to the next thought. This was not restricted to a particular speaker, but occurred widely and was quite noticeable in the way many of the participants spoke. This did not seem to be due to a lack of vocabulary in English but more to be a typical style of conversation.

The Research Process

Interviews as research instruments have been constructed within the discourse of Western academic research, and traditionally have not been familiar tools of research within the Indian culture. For that matter, the process of formal research itself has largely originated in the West, starting first with positivistic and scientific research. Interview-based studies are more often dependent on the honesty of the participants. Here too my prior experiences in school settings similar to those of the participants, in addition to the virtue of having been a native of New Delhi, brought a higher level of ease and comfort to the interview sessions. Emotions are a complex and ambiguous domain, and although Indian people in general are used to expressing their opinions, they are not used to analyzing and categorizing their feelings the way Westerners commonly do under the strong influence of psychoanalysis. I was acutely aware of this dynamic and was able to ask appropriate and probing questions during the course of the interviews such that the participants felt comfortable in reflecting candidly upon their experiences and sharing their thoughts with me.

Despite having lived in Delhi for most of my life, I was entering this research process as the outsider who had come here to gather data. In my role as a qualitative, interpretive, and naturalistic researcher, I found myself immediately taking on that mantle and proceeding to make mental notes; classifying, categorizing, and making sense of incidents, behaviors, words, thoughts, and attitudes that came at me from all directions; reading local magazines and newspapers; viewing local television; having conversations with participants in my research as well as with nonparticipants. All in an effort to soak in the sociocultural–political milieu of Delhi as would an outsider. This research was defined not only by the actual participants in the study but also by other people, places, and things that surrounded me, the researcher, and the research process in New Delhi. I wrote down descriptions not only of school observations and classroom visits, but also of other experiences—family gatherings, invitations, religious ceremonies, historical tours, conversations with friends. Wearing the lens of an outsider, I attempted to objectively pick up on the smallest details in the sociocultural–political–historical texture of the tapestry in which this qualitative study was situated. I hastily scribbled notes on a notepad every time a thought occurred to me, and every time someone did or said something that connected to something in my meaning-making process and helped fit one more piece into the complex patterns that were beginning to emerge in the research data.

Keeping a research journal proved to be invaluable. Months, or even weeks later, I began to notice that the clarity of thoughts and feelings, and the power of detail of the sounds and sights experienced in the field dimmed and began to fade. During the process of data analysis this journal provided me with valuable access to those specific feelings and thoughts that eventually contributed to deeper insights into the perceptions of the participants. The decision to transcribe all the interviews myself personally led to an intimacy between the data and myself as the researcher. Even while I was still transcribing the interviews, I had begun to engage with the text and an informal and open analysis of the data was already mentally under way in my mind. The process thereafter began to become increasingly murky and confusing as I found myself surrounded by countless sheets of interview transcripts. However, I decided to start with a closed analysis of the data by tightly focusing on the specific research questions, and began the process of creating categories as I tagged all the units of texts that went into each of the research questions. Having done that I then began to form subcategories while also looking for unexpected findings through a process of open analysis of the data. A question I constantly asked myself was: "What does this remind me of" and responded to this question through the use of any images, metaphors, and analogies that would pop into my mind.

Meaning-Making with Many "Selves"

During the various mental dialogues that I had begun to have with myself in the asking of the above question, several identities emerged in my consciousness as I began to hear the voices of my many discreet "selves," which were culturally separate but coexisting within me.

There was the *Researcher Self,* that part of me which was engaged in the actual research in the form of active field work as I interviewed participants; designed and planned for the research procedures; wrote field and observational notes; gathered data from document review; reflected, questioned, and made note of connections between various pieces of data.

There was the *Historical Self,* the voice of prior experiences and distinct memories from having been where the participants were—inside the classrooms of the teacher education colleges and the classrooms of these private schools in New Delhi; of having lived and grown up in Delhi with my immediate and extended family; and just like many of the participants, of having raised my own children in Delhi when they were young.

There was the *Indian Self* that was rooted deeply in the Indian culture and its values, beliefs, religion, philosophy, mythology, folklore, and all the multiple realities representing all of the sociocultural experiences that are incorporated in the discussion of this book.

More specifically, there was the *Familial Self,* where even as a researcher and note-taker, I still had to be supremely aware of my duties and responsibilities in my various roles among all my family members—as a daughter, as a grand-daughter, as a mother, as a sister, as a sister-in-law, as a niece, as an aunt. Being a researcher, I was acutely aware of these dynamics, but I also noted in wonder that although each role had its own expectations and demands I did not, at any point, feel a sense of encroachment but rather carried out these duties willingly and happily. It was surprising how time seemed to almost stretch to fit all the activities into a 24-hour long day.

There was also the *Colonized Self*, that part of my upbringing and education which reflected how extensively colonization had been pervasive in India: in my flawless English; in my twelve years of schooling in a Catholic Convent; in my preference for reading the works of Thomas Hardy rather than of Tagore and Gandhi; in my ingrained belief that speaking in English would elevate a person into someone more superior to those who spoke in Hindi; that the picture perfect image was a pretty cottage in the English countryside, with a white picket fence, and lilac lace curtains fluttering in kitchen windows (an image that for me only existed in the pages of the English novel and was nowhere to be found in the India that I knew); and in the Eurocentric curriculum I had studied in school and that still lingered in clear memories of my history, math and English text books. I don't remember a single text I had as a schoolgirl that wasn't written by English authors, wasn't published in England, and was not located within the English-speaking world, except of course for the Hindi texts.

There was the *American Self*, which had experienced the liberating sense of the American culture after I had immigrated to the United States. It had been exciting to study and work here, and experience the recognition and promotion for a job well done. My career really had begun here and has continued to advance because of the freedom that academic and professional opportunities have allowed me in being able to question, challenge, and explore issues that have been closest to my heart, specifically those issues that stand to be pivotal in the relationship between a society's education and its deeper values and beliefs.

There was the *Postcolonial Self*, representing the marginalized voice speaking from within the realms of Western academia, the voice of a female scholar of color questioning the discourse of mainstream early childhood education in the West, trying to find a space to make a statement and an opportunity to be heard, attempting to bring out into an open forum the realities of early childhood education in a non-white, non-Western culture through the voices of teachers in India. It was the postcolonial part of me wanting to make sense of the continued transactions between ideas and practices that were Indian and Western, traditional and modern, colonized and colonizing within my own mind while also recognizing that this psyche represented a wider Indian psyche as well.

And overarching all this was the *Speculator Self*, the objective voice within me as the researcher that identified these conflicting, dialoging, coexisting selves and made the connections between the various voices while situating them in a common matrix.

It was through the multiple dialogues between these voices of various "selves" that enabled me as the researcher to get to the core of my inquiry, to be able to demystify Indian philosophy and see the strong connections between its seemingly abstract existential ideas and the concrete reality with which those ideas played out in the day-to-day lives of so many children and adults in India. During the process of conducting and writing this research, I was able to reach a level of understanding that illustrated to me how an ancient system of beliefs conversed and dialogued with newer ideas from the West, and shaped the way in which teachers and children in India learnt, interacted, and constructed knowledge, and how the teachers themselves defined educational objectives for the children.

Insider–Outsider Perspectives

Theory building is a complex and nonlinear process, and the researcher has to draw from other theories, her own and others' insights, and empirical evidence (Graue & Walsh, 1998) Peshkin's concept of subjectivity with a focus within the researcher's own personal history, in this case mine, was very evident during the research process (Peshkin, 1988). According to Graue and Walsh (1998), insights and hunches are contextualized and emerge from one's own experiences, others' experiences, readings and con-versations, and other theories. However, the researcher also has to establish connections with the larger world of similarly held views and theoretical considerations in the field.

The background and prior experience of myself as the researcher placed me in a unique position: the "insider" and the "outsider" from different perspectives. By virtue of having spent several years as an early childhood classroom teacher and administrator in a private school in New Delhi, I was certainly no stranger to the context of this study and the issues that were being examined. Although I was an "outsider" in being a researcher from the world of American academia, I was still viewed as an "insider" from the perspective of my prior experience. I was familiar with the local discourse of early childhood education in New Delhi, and was able to enter the frames of reference and relevance of the participants more easily, and contextualize their perceptions in a more accurate manner.

On the other hand, I had also spent ten years as an early childhood classroom teacher and administrator in a progressive independent school in New York City, and had simultaneously been closely affiliated to a college of education as a student as well as an instructor. Although this made me an "outsider" to the participants of this study, it nevertheless gave me an "insider's" perspective on the debates within the early childhood discourse in the West. This unique and dual positioning of the "researcher" also served to bring me face to face with the cultural tensions between the "insider" and "outsider" perspectives on early childhood education.

There was constant negotiation between the researcher as the "insider" and the researcher as the "outsider." Needless to say, both came with prejudices, the former biased with a lens tinted by Indian beliefs and values, and the latter biased with the principles and values of American academic research. These biases could not be ignored, and the best that could be done was for me to keep an open mind, not be hasty to judge either perspective, and thus maintain a balanced approach during data analysis. While journeying through the realm of data analysis and reporting the findings, I traveled back and forth between the findings, theoretical frameworks, and the concerns raised in earlier chapters.

The Postcolonial Challenge

It can be reasonably said that education as a systemic phenomenon in India is complex, multilayered, and confusing, and the early childhood curriculum I observed did not fit into any single identifiable Western model or educational philosophy. This assumption was definitely true of the image of early childhood education that had emerged in the context of my study. After trying to tease the elements apart, three distinct strands in early childhood education in private schools in New Delhi were identified, closely intertwined with each other. These represented the discourse of *Indian Philosophy and Culture* that was seen to influence somewhat both content of curriculum and methodology of teaching in the hidden curriculum and informal teaching and learning that happened in the classroom; the discourse of *British Colonialism* that also seemed to influence content and methodology, and is manifested in the rigidly structured tests, exams, work books, curriculum, drilling, teaching to the exam; and finally, the discourse of *American Progressive Education* that appeared somewhat sporadically in the form of activity-based teaching, child-centered classrooms, desire for small class-size, and thematic teaching, but which was restricted mostly to early childhood classrooms in private schools.

This study provided an arena for early childhood teachers in New Delhi to be heard in both intensive and extensive ways. Their stories, experiences, and reasoning shed much light upon understanding early childhood practice in New Delhi, and enabled me to contextualize that practice within the Indian culture where it was taking place. I had assumed that the early childhood teachers in urban India would be desirous of simulating Western early childhood practice. This was not the case, however. The teachers did, during their practice, step into a gray area where they accepted Western practice as the dominant and standard discourse, but where they also struggled to make sense of it in relation to Indian culture. All three teachers who were interviewed seemed to come out of that quandary very quickly by realizing that the dominant discourse may be viewed as the standard in the West but it could not work in the Indian situation, and that any Western technique would have to be first modified for the cultural, emotional, physical, and academic climate of the Indian classroom. This was a good example of how the postcolonial situation is reflected in the manner in which the teachers as individuals constantly negotiate between Western and Indian ideas.

Many participants spoke about children today becoming more "Western," and the teachers made note of the strong influence of Western media and the toy industry on the children in their classrooms, appearing in conversations amongst children that focused on superhero toys and video games. This, then, brings the discussion to the conceptual implication of colonization, and an extension of neocolonialism (Altbach, 1971). Does colonization occur only after the occupation of physical and geographical territory? Is colonization a thing of the past? Indian children tend to become conditioned to a Western worldview through the Western occupation of cyberspace, media, entertainment, and other avenues used for globalization. Instead of political colonization, as has been the case historically, technologically advanced nations are now engaged in establishing economic and financial colonies. This is evident in the changing landscape of urban India: in the rapid mushrooming of McDonalds and Pizza Huts; construction of glitzy malls and cineplexes; establishment of ATM banking; Gap and Levi stores in the marketplaces juxtaposed with the Indian sari and jewelry shops; stacks of *Childhood Education* and other similar journals in the libraries of private schools; teaching students songs from the PBS television's "Barney" show and from Disney's version of The Lion King; MTV-influenced ITV channel on cable television in New Delhi; and making of the Indian versions of American television shows such as Sesame Street, Who Wants to Be a Millionaire, and American Idol. Even the theme park concept in the United States is rapidly entering the milieu of Indian society, and recently I was fascinated to read about plans to construct, near the Indian holy city of Hardwar, a new "spiritual" theme park, which will center around stories and characters from Hindu mythology. These are some examples of the juxtaposition of Indian and Western ideas that is so pervasive in the daily lives of Indians. Indian culture seems to assimilate elements, never throwing anything away, and for anyone standing on any street corner in urban India it is easy to see "one culture getting under another's skin" in the surrounding signs and languages (Iyer, 2004).

The School, as most other private schools, is one that has been constructed within this specific urban, middle-class, cultural context, which has a very strong interaction between Indian and Western practices, where Western includes both colonial and progressive educational practices. The teacher in this context continues to balance values underlying the traditional philosophy; the values underlying a textbook-driven, exam-oriented curriculum; the values underlying a child-centered classroom; as well as the values of Western lifestyles as depicted by the media in all its forms. Keeping the larger culture in mind, it is easy to understand that on a given morning a teacher such as Vasudha can incorporate in her practice, quite effortlessly, the teaching of values, academic work, and a circle time, where each of these is based on a different philosophical and pedagogical approach. This early childhood teacher negotiates a balance between tradition and the so-called 'progressivism,' and the balance between the Indian, colonial, and postcolonial elements in her practice. Her voice as the "other," of the so-called 'marginalized,' the non-Western, the old-fashioned, the unscientific, is in dialogue with the voice of the Western dominant discourse that is appealingly rational, scientific, psychological, and modern. It must be emphasized that the dominant discourse in the Indian early childhood teacher's educational practice is not the dominant discourse heard in the West, but one that is hers alone and couched in the language and values of the local community. However, for the Indian teacher this is not a battle between a right and a wrong pedagogy and elements from both can be incorporated, as we have seen. A hybrid classroom practice is thus created that is the result of a mix of different discourses. This cultural hybridity offers a new space, one that can be used for negotiation of meaning and representation (Bhabha, 1994). As I have already stated earlier, the discourse of urban Indian early childhood education and the nature of this complex curriculum defies definition and can only be understood within the locus of the system of intersecting values and beliefs that is constantly evolving, and profoundly influencing daily life inside and outside the Indian classroom.

The very concepts of "discourse," "research," "researcher," and "postcolonial" are Western constructs, and postcolonial theory speaks in the voice of Western academia (Dhillon, 1997). This study attempts to give space for the voice of the "marginalized" early childhood educator to be heard, as I attempted to bring the voice speaking a non-Western discourse into the cacophonic arena of the dominant early childhood discourse of the West. But at the same time, I found myself in an ambiguous dilemma.

In so doing, does it position the two discourses to reinforce the disparity and conflict between "acceptable" and "nonacceptable" early childhood practice? I did struggle to decide which was worse—to leave the Indian early childhood practice where it is—in a context where it occupies the position of the dominant perspective—or to bring the voice of the urban Indian early childhood teacher into the mainstream of Western discourse and run the risk of, perhaps, tagging it even more distinctly as the "marginalized" voice? Another issue that I found problematic in bringing the Indian early childhood teacher into the realm of Western research was the potential "professionalization" of the Indian teacher similar to the way that teachers are trained in the United States as they are taken through an intricate system of degrees in teaching, certifications, and licensing. Would the "whole" Indian teacher, after being thoroughly researched, run the risk of being split into a personal self and a professional self and forced to exist separately, the former out of school and the latter within the walls of the classroom, as does many a teacher in the West? Does this kind of research help or compound the problem? Attempts to professionalize the early childhood teacher seemed also to dehumanize her in some ways. And yet I was nagged by the fact that the knowledge that is "heard" most is that which is presented by Western research, and if the "marginalized" is not brought into the domain of formal research it will tend to remain marginalized. If research focuses only on Western subjects/issues/concepts then only Western subjects/issues/concepts will continue to be prioritized and valued as sources of universal truths. The field of educational research would benefit through the inclusion of topics that discuss perspectives on the educational practice of the non-West, allowing the knowledge matrix to resemble cultural truths in a clearer and more accurate way.

The study upon which this book is based reveals an educational system that is a hybridization of beliefs, practices, and pedagogies. There are many instances as illustrated through the voices of the participants that indicate levels of comfort as well as discomfort as they negotiate different sets of beliefs, values, and life styles not only in their daily living but also in their classrooms. This particular study was based in urban private schools within a certain socioeconomic population. Again, I would like to remind the reader that there is great diversity in the context of schooling for young children in India, and studies of schools and classrooms that are government-run, or are situated in rural or poorer areas of the nation would present a very different picture. It is my hope that this book will enable teachers to identify and name dimensions of the processes they are involved in, and ask questions such as: Why do classrooms and educational philosophies look the way they do? How do we explain the need for schools in India to teach English, Hindi, and a regional language simultaneously? How do we explain the need to celebrate and close schools for ten different religious holidays? How do we make sense of what we are teaching in the ways in which we are teaching? What images form the focus of our stories? Are these images Hindu, Muslim, Christian, Buddhist, Sikh, or Jain? Are any socio-cultural images excluded from the curriculum? Are our stories coming from local texts or are they remote Euro-American stories with colonial messages? How do we understand the need for stories from both these sources? In all the above will be embedded a space within which the same colonizing questions can be asked of Indian educators and policymakers themselves: How do years of a caste ideology where upper-caste children have been privileged and girls have been excluded impact the ways in which schools support the education of its children? Deliberately and persistently examining one's own practice and confronting such questions leads to a deeper understanding of how the ways of being in a classroom are situated within larger educational and political contexts. It is also my hope that such studies will not only facilitate educators' recognition of a world where there is an increasing need to formally balance Vygotsky and the *Veda*, but also empower teachers who are familiar with different discourses to contact each other and begin a dialogue that would work toward increasing the accountability of all educators everywhere.

References

Achinstein, B., Ogawa, R.T., & Speiglman, A. (2004). Are we creating separate and unequal tracks of teachers? The effects of state, policy, local conditions, and teacher characteristics on new teacher socialization. *American Educational Research Journal*, 41(3), 557–603.

Achyuthan, M. (1974). *Educational practices in Manu, Panini, and Kautilya*. Trivandrum, India: M. Easwaran Publishers, College Book House.

Alexander, R.J. (2000). *Culture and pedagogy: International comparisons in primary education*. Oxford, UK: Blackwell.

Allsup, C. (1995). What's all this about male bashing? In R.J. Martin (Ed.), *Practicing what we teach: Confronting diversity in teacher education*. Albany, NY: State University of New York Press.

Altbach, P.G. (1971). Education and neocolonialism: A note. *Comparative Education Review*, 15, 237–239.

Altekar, A.S. (1965). *Education in ancient India*. Varanasi, India: Nand Kishore & Bros.

Alva, J.J.K. de. (1995). The Postcolonization of the Latin American experience: A reconsideration of "Colonialism," "Postcolonialism" and "Mestijaze." In G. Prakash (Ed.), *After colonialism, imperial histories, and postcolonial displacements* (pp. 241–275). Princeton, NJ: Princeton University Press.

Bandura, A. (1997). *Self-efficacy: The exercise of control*. New York, NY: Freeman.

Bartolome, L. (1994). Beyond the methods fetish. *Harvard Educational Review*, 64(2), 173–194.

Basham, A.L. (1998). *The wonder that was India*. New Delhi, India: Rupa & Co.

Bernard, T. (1995). *Hindu philosophy*. Bombay, India: Jaico Publishing House.

Bernhard, J.K., Gonzalez-Mena, J., Chang, H.N., O'Loughlin, M., Eggers-Pierola, C., Roberts Fiati, G., & Corson, P. (1998). Recognizing the centrality of cultural diversity and racial equity: Beginning a discussion and critical reflection on "Developmentally Appropriate Practice." *Canadian Journal of Research in Early Childhood Education*, 7(1), 81–90.

Bhabha, H. (1994). *The location of culture*. London, UK: Routledge.

Bhagvad Gita. Trans. E. Easwaran (1985). California: Nilgiri Press.

Blatchford, P., Moriarty, V., Edmonds, S., & Martin, C. (2002). Relationships between class size and teaching: A multimethod analysis of English infant schools. *American Educational Research Journal*, 39(1), 101–132.

Bloch, M.N. (1992). Critical perspectives on the historical relationship between child development and early childhood research. In S.A. Kessler & B.B. Swadener (Eds.),

Reconceptualizing the early childhood curriculum: Beginning the dialogue (pp. 3–20). New York, NY: Teachers College Press.

Bloom, B.S. (1956). *Taxonomy of educational objectives, Handbook 1: The cognitive domain.* New York, NY: David Mckay Co. Inc.

Bowman, B.T. & Stott, F.M. (1994). Understanding development in a cultural context: The challenge for teachers. In B.L. Mallory & R.S. New (Eds.), *Diversity & developmentally appropriate practices: challenges for early childhood education* (pp. 119–133). New York, NY: Teachers College Press.

Bredekamp, S. (1987). *Developmentally appropriate practice in early childhood programs serving children from birth through age 8.* Washington, DC: National Association of Education of Young Children.

Bredekamp, S. & Copple, C. (Eds.) (1997). *Developmentally appropriate practice in early childhood programs.* Rev. ed. Washington, DC: National Association for the Education of Young Children.

Britzman, D., Dippo, D., Searle, D., and Pitt, A. (1997). Toward an academic framework for thinking about teacher education. *Teaching Education Journal,* 9(1), 15–26.

Brouwer, N. & Korthagen, F. (2005). Can teacher education make a difference? *American Educational Research Journal,* 42(1), 153–224.

Burbules, N. (1997). A grammar of difference: Some ways of rethinking difference and diversity as educational topics. *Australian Educational Researcher,* 24(1), 97–116.

Campbell, R.J., Kyriakides, L., Mujis, R.D., & Robinson, W. (2003). Differential teacher effectiveness: Towards a model for research and teacher appraisal. *Oxford Review of Education,* 29(3), 347–362.

Canella, G.S. (1997). *Deconstructing early childhood education: Social justice & revolution.* New York, NY: Peter Lang Publishing, Inc.

Chandra, A.N. (1980). *The Rig Vedic culture and the Indus civilization.* Calcutta, India: Ratna Prakashan.

Chester, M. & Beaudin, B. (1996). Efficacy beliefs of newly hired teachers in urban schools. *American Educational Research Journal,* 33(1), 233–257.

Clandinin, D.J. (1986). *Classroom practice: Teacher images in action.* London: The Falmer Press.

Clandinin, D.J. & Connelly, F.M. (1986). Rhythms in teaching: The narrative study of teachers' personal practical knowledge of classrooms. *Teaching and Teacher Education,* 2, 377–387.

Clandinin, D.J. & Connelly, F.M. (1992). Teacher as curriculum maker. In P.W. Jackson (Ed.), *Handbook of research on curriculum* (pp. 363–401). New York, NY: Macmillan Publishing Company.

Clark, C.M. & Peterson, P.L. (1986). Teachers' thought processes. In M. Wittrock (Ed.), *Handbook on Research on Teaching.* New York, NY: Macmillan Publishing Company.

Cremin, L.A. (1988). American education: The metropolitan experience 1876–1980. New York, NY: Harper & Row.

Dave, I. (1991). *Indian personality in its developmental background.* Udaipur, India: Himanshu Publications.

Davies, S. (2002). The paradox of progressive education: A frame analysis. *Sociology of Education*, 75(4), 269–286. Available at http://www.links.jstor.org/sici.

de Nicolas, Antonio T. (1976). *Avatara: The humanization of philosophy through the Bhagvad Gita*. New York, NY: Nicolas Hays.

Delpit, L. (1995). *Other people's children: Cultural conflict in the classroom*. New York, NY: The New Press.

Desperately seeking another Vivekananda (2002, January 12). New Delhi: *The Times of India*, p. 12.

Dewey, J. (1933). *How we think*. Boston, MA: Heath.

Dhillon, P.A. (1997). *Abhijnana: The forgetting and re-membering of Indian Thought*. Online journal available at www.ed.uiuc.edu/EPS/MWCIES97/dhillon

di Bona, J. (1983). *One teacher, one school*. New Delhi, India: Biblia Impex Pvt. Ltd.

Dube, S.C. (1990/2000). *Indian society*. New Delhi, India: National Book Trust.

Education is key to spirituality (2002, January 7). New Delhi: *The Times of India*, p. 12.

Elbaz, F. (1981). The teacher's practical knowledge: Report of a case study. *Curriculum Inquiry*, 11, 43–71.

Ellinger, H. (1995). *The basics: Hinduism*. PA: Trinity Press International.

Farber, K.S. (1995). Teaching about diversity through reflectivity: Sites of uncertainty, risk, and possibility. In R.J. Martin (Ed.), *Practicing what we teach: Confronting diversity in Teacher Education*. Albany, NY: State University of New York Press.

Feiman-Nemser, S. (2001). From preparation to practice: Designing a continuum to strengthen and sustain teaching. *Teachers College Record*, 103, 1013–1055.

Fosnot, C.T. (1996). Constructivism: A psychological theory of learning in C. Fosnot (Ed.), *Constructivism: Theory, perspectives, and practice*. New York, NY: Teachers College Press.

Friedman, T.L. (August 14, 2002). Where freedom reigns (Op-ed: Week in Review). *The New York Times*, p. 13.

Gandhi, L. (1998). *Postcolonial theory: a critical introduction*. New York, NY: Columbia University Press.

Gandhi, M.K. (1938). *Hind Swaraj*. Reprint. Ahmedabad, India: Navjivan Publishing House.

Gay, G. (1995). Modeling and mentoring in urban teacher preparation. *Education and Urban Society*, 28(1), 103–118.

Geertz, C. (1973). *The interpretation of cultures*. New York, NY: Basic Books.

Giroux, H.A. (1989). *Teachers as intellectuals*. MA: Bergin & Garvey.

Gokhale, N. (2005). Educating a community to educate their young. *International Journal of Early Childhood*, 37(2), 21–28.

Gotz, I.L. (1995). Education and the Self: Cross-cultural perspectives. *Educational Theory on the Web*, 45(4). Available at www.ed.uiuc.edu/EPS/Educational-Theory/Contents/45_4

Goyal, P. (2003). Education in pre-British India. Posted on 7/25/2003 at www.infinityfoundation.com

Grant, C.A. & Wieczorek, K. (2000). Teacher education and knowledge in "the knowledge society: The need for social moorings in our multicultural schools." *Teachers College Record*, 102(5), 913–935.

Graue, M.E. & Walsh, D.J. (1998). Theory as context. In M.E. Graue & D.J. Walsh (Eds.), *Studying children in context: Theories, methods, and ethics* (pp. 24–54). Thousand Oaks, CA: Sage.

Grieshaber, S. & Canella, G.S. (2001). From identity to identities: Increasing possibilities in early childhood education. In S. Greishaber & G. Canella (Eds.), *Embracing identities in early childhood education: diversity and possibilities* (pp. 3–22). New York, NY: Teachers College Press.

Gupta, A. (2001). Implementing change at the pre-primary level in a school in India. *International Journal of Early Childhood*, 33(1), 34–42.

Gupta, A. (2003). Socio-cultural-historical constructivism in the preparation and practice of early childhood teachers in New Delhi, India. *Journal of Early Childhood Teacher Education*, 24(3), 163–170.

Gupta, A. & Raman, U. (1994). *Roundabout A: Basic conceptual skills*. New Delhi, India: Orient Longman Limited.

Gupta, A. & Raman, U. (1994). *Roundabout B: Language and number skills*. New Delhi, India: Orient Longman Limited.

Guyton, E. & Hidalgo, F. (1995). Characteristics, responsibilities, and qualities of urban school mentors. *Education and Urban Society*, 28(1), 40–47.

Hollingsworth, S. (1989). Prior beliefs and cognitive change in learning to teach. *American Educational Research Journal*, 26(2), 160–189.

hooks, b. (1989). *Talking back: Thinking feminist, thinking Black*. Boston, MA: South End Press.

Howe, K. & Berv, J. (2000). Constructing constructivism, epistemological and pedagogical. In D.C. Phillips (Ed.), *Constructivism in Education*. Chicago, IL: University of Chicago Press.

Husain, Abid, S. (1978). *The national culture of India*. New Delhi, India: National Book Trust. India's poor bet precious sums on private schools (2003, November 15). New York: *The New York Times*, section A, p. 1.

India "world's biggest readers" (2005, June 27). BBC News. http://news.bbc.co.uk/go/pr.

Iyer, P. (2004). *Sun after dark: Flights into the foreign*. New York, NY: Alfred. A. Knopf.

Jalongo, M.R. & Isenberg, J.P. (1995). *Teachers' stories*. San Francisco, CA: Jossey-Bass, Inc. Publishers.

Jipson, J. (1991). Developmentally appropriate practice: Culture, curriculum, connections. *Early Education and Development*, 2(2), 120–136.

Kakar, S. (1981). *The inner world: A psycho-analytic study of childhood and society in India*. New Delhi, India: Oxford University Press.

Kalam, Abdul, A.P.J. (2002). *Ignited minds: Unleashing the power within India*. New Delhi, India: Viking by Penguin Books.

Kamii, C. (1984). *Young children reinvent arithmetic: Implications of Piaget's theory*. Washington, DC: National Association for the Education of Young Children.

Kapoor, S. (2002). *Nai Leher Book Series, Part 3*. New Delhi, India: Oxford University Press.

Katz, L.G. (1996). Child development knowledge and teacher preparation: Confronting assumptions. *Early Childhood Research Quarterly*, 11(2), 135–146.

Katz, L.G. (1999). International perspectives on early childhood education: Lessons from my travels. *Early Childhood Research and Practice*, 1(1). Available at http://eecrp.uiuc.edu/v1n1/katz.html.

Katz, R. C. (1989). *Arjuna in the Mahabharata*. Columbia, SC: University of South Carolina.

Kaul, V. (1998, March). *Minimum standards for quality in early childhood education*. Paper presented at a meeting on The National Consultation Meet on Streamlining of Early Childhood Education Services. New Delhi, India.

Keay, F.E. (1918/1980). *Ancient Indian education: An inquiry into its origin, development, and ideals*. New Delhi, India: Cosmo Publications.

Keay, J. (2000). *India: A History*. New York, NY: Grove Press.

Kennedy, D. (2000). The roots of child study: Philosophy, history, and religion. *Teachers College Record*, 102(3), 514–538.

Kindling the Vedic spirit in the age of Dotcom (2000, July 30). New Delhi: *The Times of India*, p. 6.

King, N. (1979). Play: The kindergartners' perspective. *The Elementary School Journal*, 80(2), 81–87.

Kliebard, H.M. (1995). *The struggle for the American curriculum 1893–1958*. New York, NY: Routledge.

Kohlberg, L. (1969). Stage and sequence. In D. Goslin (Ed.), *Handbook of socialization theory and research*. Chicago, IL: Rand MacNally.

Korthagen, F.A.J. & Kessels, J.P.A.M. (1999). Linking theory and practice: Changing the pedagogy of teacher education. *Educational Researcher*, 28 (4), 4–17.

Krishnamurti, J. (1974). *Krishnamurti on education*. London, UK: Krishnamurti Foundation Trust. Ltd.

Kumar, K. (1992). *What is worth teaching?* New Delhi: Orient Longman Limited.

Kumar, K. (1993). Literacy and primary education in India. In P. Freebody and A. Welch (Eds.), *Knowledge, culture and power: International perspectives on literacy as policy and practice* (pp. 102–113). London: Falmer Press.

Ladson-Billings, G. (1995). Toward a theory of culturally relevant pedagogy. *American Educational Research Journal*, 32(3), 465–491.

Lampert, M. (1985). How do teachers manage to teach? Perspectives on problems in practice. *Harvard Educational Review*, 55(2), 178–194.

Mahabharata. Trans. P.C. Roy. Calcutta, India: Bharat Press, 1884–1896.

Mahmood, S. (1895). *A history of English education in India. Its rise, development, progress, present condition and prospects being a narrative of the various phases of educational policy and measures adopted under the British Rule from its beginning to the present period (1781–1893)*. Delhi, India: Idarah-I Adabiyat-I Delli.

Making changes, making knowledge: Living, learning and teaching together. Panel presentation by Franklin, C., Garavuso, V., Gupta, A., Malone, C., Miletta, A., & Wilgus, G. at AERA, Montreal, April 14, 2005.

Marso, R. & Pigge, F. (1997). A longitudinal study of persisting and non-persisting teachers' academic and personal characteristics. *Journal of Experimental Education*, 65(3), 243–254.

Martin, R.J. (1995). Deconstructing myth, reconstructing reality: Transcending the crisis in teacher education. In R.J. Martin (Ed.), *Practicing what we teach: Confronting diversity in teacher education.* Albany, NY: State University of New York Press.

McAllister, G. & Irvine, J.J. (2000). Cross cultural competency and multicultural teacher education. *Review of Educational Research, 70*(1), 3–24.

McLaren, P. (1994). *Life in schools: An introduction to critical pedagogy in the foundations of education.* New York, NY: Longman.

Mehta, G. (1997). *Snakes and Ladders: Glimpses of modern India.* London, Great Britain: Vintage-Random House.

Miles, M. (1997). Disabled learners in South Asia: Lessons from the past for educational exporters. *International Journal of Disability, Development and Education, 44* (2), 97–104.

Mills, C.J. (2003). Characteristics of effective teachers of gifted students: Teacher background and personality styles of students. *Gifted Child Quarterly, 47*(4), 272–281.

Moll, L. (1990). *Vygotsky and education: Instructional implications and applications of sociohistorical psychology.* Cambridge: Cambridge University Press.

Moll, L. & Arnot-Hopffer, E. (2005). Sociocultural competence in teacher education. *Journal of Teacher Education, 56*(3), 242–247.

Molnar, A., Smith, P., Zahorik, J., Palmer, A., Halbach, A., & Ehrle, K. (1999). Evaluating the SAGE Program: A pilot program in targeted pupil-teacher reduction in Wisconsin. *Educational Evaluation and Policy Analysis, 21*(2), 165–177.

Nanavaty, J. J. (1973). *Educational thought. Volume 1.* Poona, India: Joshi and Lokhande Prakashan.

Nandy, A. (1983). *The intimate enemy: Loss and recovery of self under colonialism.* Delhi, India: Oxford University Press.

National Policy on Education available on Department of Education, Government of India home page at www.education.nic.in.

NCERT (2001). National Curriculum Framework for School Education.

NCERT (August 2002). Report on the national conference on early childhood care and education in the context of Sarva Shiksha Abhiyan.

Need to tap Vedic knowledge (2000, July 17). New Delhi: *The Times of India: Education Times,* p. 1.

Nehru, J. (1946/1991). *The discovery of India.* India: Jawaharlal Nehru Memorial Fund: Oxford University Press.

New, R.S. (1999). An integrated early childhood curriculum: Moving from the *what* and the *how* to the *why*. In C. Seefeldt (Ed.), *The early childhood curriculum: Current findings in theory and practice* (pp. 265–287). New York, NY: Teachers College Press.

New, R.S. & Mallory, B.S. (1994). Introduction: The ethic of inclusion. In R.S. New and B.S. Mallory (Eds.), *Diversity and developmentally appropriate practice: Challenges for early childhood education* (pp. 1–14). New York, NY: Teachers College Press.

Olson, J.K. (1981). Teacher influence in the classroom. *Instructional Science, 10,* 259–275.

Ozmon, H. & Craver, S. (1995). *Philosophical foundations of education.* New Jersey: Prentice-Hall.

Paine, J. (1998). *Father India*. New York, NY: Harper Collins Publishers.

Paranjoti, V. (1969). *East and west in Indian education*. Lucknow, India: Lucknow Publishing House.

Peshkin, A. (1988). In search of subjectivity—one's own. *Educational Researcher*, 17(7), 17–22.

Phelan, A.M. (1997). In their prime: Reading the lives of non-traditional beginning teachers. *Teaching Education Journal*, 9 (1), 143–158.

Piaget, J. (1965/1932). *The moral judgement of the child*. New York, NY: Free Press.

Pratt, M.L. (1992). *Imperial eyes: Travel writing and transculturation*. New York and London. Routledge.

Ramayana of Valmiki. Trans. H.P. Shastri (1962). London: Shantisadan.

Richardson, V. (1994). Conducting research on practice. *Educational Researcher*, 23(5), 5–10.

The Rig Veda: An anthology. Trans. W.D. O'Flaherty (1994). India: Penguin Books.

Rogoff, B. (1990). *Apprenticeship in thinking: Cognitive development in social context*. New York, NY: Oxford University Press.

Rogoff, B., Mosier, C., Mistry, J.J., & Goncu, A. (1993). Toddlers' guided participation with their caregivers in cultural activity. In E.A. Forman, N. Minick, & C. Stone (Eds.), *Context for learning: Sociocultural dynamics in children's development* (pp. 230–253). Oxford, UK: Oxford University Press.

Sachs, S.K. (2004). Evaluation of teacher attributes as predictors of success in urban schools. *Journal of Teacher Education*, 55(2), 177–187.

Sacred Space (2002, January 11). New Delhi: *The Times of India*, p. 10.

Sacred Space (2005, January 17). New Delhi: *The Times of India*, p. 18.

Saini, A. (2000). Literacy and empowerment: An Indian scenario. *Childhood Education*. International Focus Issue, 2000.

Satya Prakash (1965). *Founders of Sciences in Ancient India*. New Delhi, India: The Research Institute of Ancient Scientific Studies.

Schon, D.A. (1983). *The reflective practitioner: How professionals think in action*. New York, NY: Basic Books.

Schon, D.A. (1987). *Educating the reflective practitioner: Towards a new design for teaching and learning in the professions*. San Francisco, CA: Jossey-Bass.

Schoonmaker, F. (2003). Growing up teaching: From personal knowledge to professional practice. New York, NY: Teachers College Press.

Schoonmaker, F. & Ryan, S. (1996). Does theory lead practice? Teachers' constructs about teaching: Top-down perspectives. *Advances in Early Education and Day Care*, 8, 117–151.

Semel, S.F. & Sadovnik, A.R. (1999). Progressive education: Lessons from the past and present. In S.F. Semel & A.R. Sadovnik (Eds.), *Schools of tomorrow, schools of today: What happened to progressive education*. New York, NY: Peter Lang.

Simon, R.F. (1992). *Teaching against the grain: Texts for a pedagogy of possibility*. New York, NY: Bergin & Garvey.

Singh, R.P. (1980). *Studies in teacher education: An overview*. New Delhi: Bahri Publications.

Slater, L. (2002, February 3). The trouble with self-esteem. New York: *The New York Times Magazine*, pp. 44–47.

Sleeter, C.E. & Grant, C.A. (1994). *Making choices for multicultural education: Five approaches to race, class, and gender in American Education.* Columbus, OH: Merrill.

Soto, L.D. (1999). The multicultural worlds of childhood in postmodern America. In C. Seefeldt (Ed.), *The early childhood curriculum: Current findings in theory and practice.* New York, NY: Teachers College Press.

Sparks-Langer, G. & Colton, A. (1991). Synthesis of research on teachers' reflective thinking. *Educational Leadership,* 48(6), 37–44.

Spodek, B. (1988). Implicit theories of early childhood teachers: Foundations for professional behavior. In B. Spodek, O.N. Saracho, & D.L. Peters (Eds.), *Professionalism and the early childhood practitioner* (pp. 161–172). New York, NY: Teachers College Press.

Srinivasan, B. & Karlan, G.R. (1997). Culturally responsive early intervention programs: Issues in India. *International Journal of Disability, Development and Education,* 44(4).

Srivastava, R.C. & Bose, K. (1978). *Theory and practice of teacher education in India.* New Delhi: Chugh Publications.

State Council of Educational Research and Training (September 1994). *Early Schooling: Problems and Perspectives.* New Delhi.

Stemler, S.E., Elliott, J.G., Grigorenko, E.L., & Sternberg, R.J. (2005). *Practical intelligence and teacher preparation.* Paper presented at AERA, Montreal, Canada, April 13, 2005.

Sternberg, R.J. & Horvath, J. (1995). A prototype view of expert teaching. *Educational Researcher,* 24(6), 9–17.

Stott, F. & Bowman, B. (1996). Child development knowledge: A slippery base for practice. *Early Childhood Research Quarterly,* 11(2), 169–183.

Stronge, J.H. (2002). Qualities of effective teachers. *Adolescence,* 37(148), 868.

Suleri, S. (1992). *The rhetoric of English India.* Chicago, IL: University of Chicago Press.

Sullivan, A. (2002, October 14). Lacking in self-esteem? Good for You! New York: *Time,* 160(16), 102.

Tauer, S.M. & Tate, P.M. (1998). Growth of reflection in teaching: Reconciling the models. *Teaching Education Journal,* 9(2), 143–153.

Tischer, R.P. & Wideen, M.F. (1990). *Research in teacher education: International perspectives.* London: The Falmer Press.

Thapa, V.J. (1994, June 15). Hop, skip, sing and learn. New Delhi: *India Today.*

Trivedi, H. (1993). *Colonial transactions: English literature and India.* Calcutta: Papyrus.

UNESCO (1972). Teacher education in Asia: a regional survey. Asian Institute for Teacher Educators. University of the Phillipines. Regional office for education in Asia, Bangkok.

Upanishad. Trans. E. Easwaran (1996). India: Penguin Books.

Varma, M. (1969). *The philosophy of Indian education.* Meerut, India: Meenakshi Prakashan.

Varma, P.K. (1999). *The great Indian middle class.* New Delhi, India: Penguin Books.

Viruru, R. (2001). *Early Childhood Education: Postcolonial perspectives from India.* New Delhi, India: Sage Publications.

Vishwanathan, G. (1995). The beginnings of English literary study in British India. In Bill Ashcroft, Gareth Griffiths, & Helen Tiffin (Eds.), *The Post-colonial studies reader*. New York: Routledge.

Vyas, K.C. (1981). Historical Background 2, The National Education after 1900. In Bhatia, Bannerjee, Datta, Bhattacharya, & Mahanta (Eds.), *Indianisation of our Education*. Calcutta, India: West Bengal Headmaster's Association.

Vyas, R.N. (1981). *Indian and Western educational psychologies and their synthesis*. Ambala Cantt., India: Associated Publishers.

Vygotsky, L.S. (1934/1987). Thinking and speech. In R.W. Reiber & A.S. Carton (Eds.), *The collected works of L.S. Vygotsky: Volume 1: Problems of general psychology*. New York, NY: Plenum.

Vygotsky, L.S. (1962). *Thought and language*. Cambridge, MA: MIT Press.

Vygotsky, L.S. (1978). *Mind in society: The development of higher psychological processes*. Cambridge, MA: Harvard University Press.

Walkerdine, V. (1984). Developmental psychology and the child-centered pedagogy: The insertion of Piaget's theory into primary school practice. In J. Henriques et al. (Eds.), *Changing the subject: Psychology, social regulation and subjectivity*. London: Methuen.

War must never be an alternative (2002, January 4). New Delhi: *Indian Express*.

Wideen, M., Mayer-Smith, J., & Moon, B. (1998). A critical analysis of the research on learning to teach: Making the case for an ecological perspective on inquiry. *Review of Educational Research*, 68, 130–178.

Wiles, J. & Bondi, J. (1989). *Curriculum development: A guide to practice*. Columbus, OH: Merrill Publishing Company.

Williams, L.R. (1991). Curriculum making in two voices: Dilemmas of inclusion in early childhood education. *Early Childhood Research Quarterly*, 6, 303–311.

Williams, L.R. (1994). Developmentally appropriate practice and cultural values: A case in point. In B.L. Mallory & R.S. New (Eds.), *Diverisy & Developmentally Appropriate Practices: Challenges for early childhood education* (pp. 155–165). New York, NY: Teachers College Press.

Williams, L.R. (1996). Does practice lead theory? Teachers' constructs about teaching: Bottom-up perspectives. *Advances in Early Education and Day Care*, 8, 153–184.

Williams, L.R. (1999). Determining the early childhood curriculum: The evolution of goals and strategies through consonance and controversy. In C. Seefeldt (Ed.), *The early childhood curriculum: Current findings in theory and practice*. New York, NY: Teachers College Press.

Woman of the year (2002, January 2). New Delhi: *The Indian Express*. Special section Focus on Education and Career Opportunities.

Wood, D.J., Bruner, J., & Ross, G. (1976). The role of tutoring in problem solving. *Journal of Child Psychology and Psychiatry*, 17, 89–100.

Yunus, S.M. (2005). Childcare practices in three Asian countries. *International Journal of Early Childhood*, 37(1), 39–56.

Zaehner, R.C. (1966). *Hinduism*. Oxford, UK: Oxford University Press.

Zeichner, K.M. & Tabachnick, B.R. (1981). Are the effects of university teacher education "washed out" by school experience? *Journal of Teacher Education*, 32, pp. 7–13.

Zigler, R.L.(1994). Reason and emotion revisited: Achilles, Arjuna and moral conduct. *Educational Theory on the Web*, 44(1). Available at www.ed.uiuc.edu/EPS/Educational-Theory/Contents/44_1/Zigler

Zurawsky, C. (2003). Class size: Counting students can count. *Research Points*, 1(2), 1–4. AERA: available at www.aera.net/pubs/rp/RPFall03ClassSize

Related Readings

Dewey, J. (1920/1957). *Reconstruction in philosophy*. Boston, MA: Beacon.

Eisner, E.W. (1991). *The enlightened eye*. New York, NY: Macmillan.

Erikson, E. (1963). *Childhood and society*. New York, NY: Norton.

Forman, E.A. (1992). Discourse, intersubjectivity, and the development of peer collaboration: A Vygotskian aproach. In L.T. Winegar & J. Valsiner (Eds.), *Children's development within social contexts* (pp. 143–159). Hillsdale, NJ: Lawrence Earlbaum Associates Publishers.

Ladson-Billings, G. (1994). *The Dreamkeepers*. San Francisco: Jossey-Bass Publishers.

Noddings, N. (1999). Longing for the sacred in schools: A conversation with Nel Noddings. *Educational Leadership*, 56(4), 28–32.

Piaget, J. (1962). *Play, dreams and imitation in childhood*. New York, NY: Norton.

Seidman, I. (1998). *Interviewing as qualitative research: A guide for researchers in education and the social sciences*. New York, NY: Teachers College Press.

Sharma, S.D. (1992). *Horizons of Indian education*. New Delhi, India: Sterling Publishers.

Siddiqui, D.A. (1997). Identifying educational values for the emerging global society. *Educational Theory on the Web*. Available at www.ed.uiuc.edu/EPS/MWCIES97.

Skinner, B.F. (1965). *Science and human behavior*. New York, NY: Free Press.

Index

Printed in the United States
By Bookmasters